Ian MacDonald

ESSENTIALS

GCSE Design & Technology
Resistant Materials
Revision Guide

Contents

Contents

Exam Boards and the Exam

The Exam Boards

In England there are three main awarding bodies. Each one has a website where you can access information:

- **AQA** (Assessment and Qualifications Alliance), www.aqa.org.uk
- **OCR** (Oxford, Cambridge and RSA Examinations), www.ocr.org.uk
- **Edexcel**, www.edexcel.org.uk

There are separate exam boards for Northern Ireland and Wales.

The specification (formerly called a syllabus) contains information about what you should be taught and how you will be assessed.

Ask your teacher about...
- the specification you're following
- your controlled assessment from 2010
- the type of examination you'll have to sit.

The Written Exam

Your written exam will take place at the end of the course and will make up 40% of your GCSE grade. The majority of exam boards set one single paper, but some might set two papers.

Some exam boards will provide you with a theme for part of the examination. This will be provided after Christmas, before you sit the examination. For example, a theme could be Children's Toys or Garden Furniture, and this will allow you to focus part of your revision time on this area. Be careful though, the paper(s) need to test the whole specification so not every question will be based on the theme.

Exams will be between one and a half and two hours in length, so you'll need to prepare thoroughly for this part of the assessment.

Examiners' Advice

It is the examiners' job to set a written examination that will test your knowledge and understanding related to designing and making with resistant materials.

As a general guide, they plan questions that will earn 1 mark a minute. So, if a question is worth 10 marks, you're expected to spend about 10 minutes on that question.

Revision Areas

Make sure that you revise the following topics:

- Timbers and manufactured boards.
- Metals – ferrous, non-ferrous and alloys.
- Plastics – thermoplastics and thermosetting plastics.
- Composite, smart and nano materials.
- Components used for fastening materials together.
- Adhesives, joints and joining methods.
- Tools and machinery for working materials.
- Ways of finishing materials.
- Manufacturing processes used in both school and industry.
- Mechanisms.
- Health and safety issues.
- Wider moral, social and environmental issues for design and technology.
- Anthropometrics and ergonomics.
- Computer aided design (CAD) and computer aided manufacture (CAM).

Sketching Equipment

It is a good idea to take the following items into the exam with you:

- Pen / ball-point pen – for writing all text (but not for sketches or diagrams).
- Graphite pencil – HB for sketching and 2B for shading.
- Coloured pencils – to show different materials and textures and to give drawn shapes three dimensions. (N.B. students taking the Edexcel exam can **only** use a black pencil for sketching).

Freehand Sketching

Technique

You should practise drawing rectangles, triangles and circles. These are the basic elements of many products. If you combine them (crating), you can draw almost anything.

Use a pencil to draw shapes in the exam, but don't waste time rubbing out mistakes. Simply cross out any incorrect answers.

Practise drawing simple ideas until you can sketch quickly and accurately for the exam.

Design Sketches

The examiner is looking for **two-dimensional (2D)** and **three-dimensional (3D)** design sketches which are...

- crisply drawn
- easy to understand
- detailed (including information such as the material(s), sizes, joints and finishes).

One way to do this is to use **isometric** or **oblique** sketches. This allows solid objects to be drawn with depth.

If you draw a sketch of an **exploded** joint, this means that the parts of the joint have been moved apart for clarity.

A **cross-sectional** view allows you to see a cut-away view.

N.B. The thickness of materials should always be drawn.

Exploded View of Joint

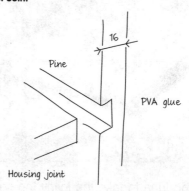

Cross-sectional View of Fitting (Hexagonal Bolt and Nut)

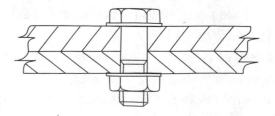

Information Sketches

Information sketches are used to give information about processes or tools.

It's best to draw them two dimensionally, so that only the important details are shown.

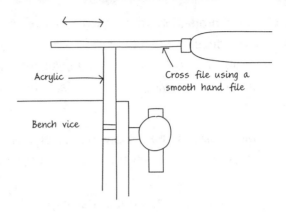

Rendered and Textured Sketches

Rendering means applying colour and shade to an object to make it look realistic. You should use rendering to show shadows on solid objects.

Texture will also give the appearance of real materials.

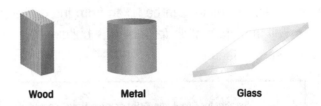

Wood   Metal   Glass

Sketch of a Garden Kneeler

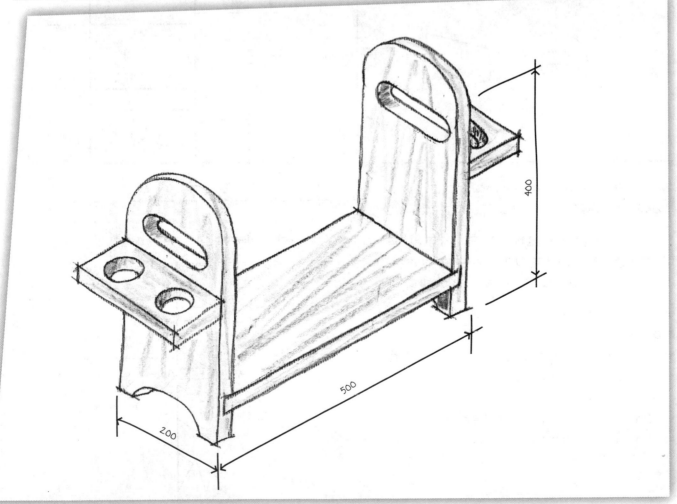

Working Drawings

Orthographic Detail Drawings

Orthographic projection drawings, also known as **working drawings**, are used in industry to provide clear information about a design. The designer will provide a detailed drawing for a manufacturer to follow.

The drawings give the necessary instructions for a product to be made. The advantages of this system are that…

- everyone can understand the drawings
- it's easy to understand all the views
- the views are accurate without any distortion
- measurements can be taken from the drawings
- all the information for making / building is in the same place.

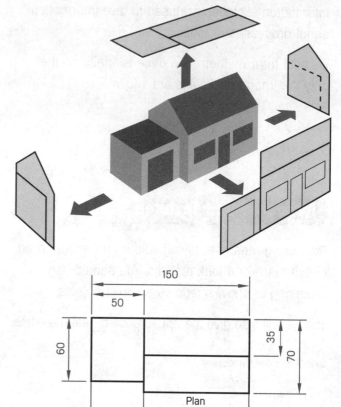

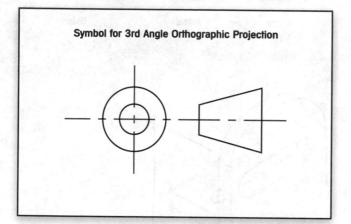

Symbol for 3rd Angle Orthographic Projection

Plan

Elevation

Orthographic Sketches

You will not have enough time in the exam to make a full orthographic drawing, but a sketched orthographic drawing can be very useful for adding details.

Orthographic drawings could include the following information:

- **Dimensions** or measurements (at least three).
- Details or enlarged cross-sections to show extra information.
- **Cross-section** views for hidden details.
- Information labels for materials, joints and finishes.

Cross-sectional View of Joint

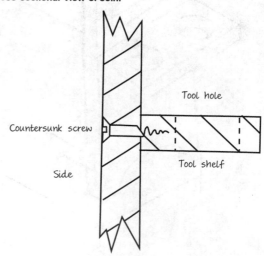

Tool hole

Countersunk screw

Tool shelf

Side

Prototype Modelling

Working in 3D can produce very different design ideas from simply sketching with paper and pencil. Designers often model to get a solid impression of an idea they have designed on paper.

You should know the suitable modelling materials, including…

- sheet cardboard
- corrugated cardboard
- corrugated plastic sheet
- polystyrene sheet
- straws
- polymorph
- styrofoam.

3D and Virtual Modelling

Scale models and **prototypes** are often made by designers.

From a design point of view there are many advantages of making models in 3D, for example…

- the design idea can be viewed from all sides
- you get a better idea of scale and proportion
- some construction problems can be identified
- the client and other users can be consulted
- modifications can be made easily.

Virtual modelling uses a computer to produce 3D images on screen.

Quick Test

1. List three advantages of using orthographic drawings.
2. Make a 2D sketch of a piece of acrylic being filed in a vice.
3. Make a 3D sketch of a housing joint. Add rendering to your sketch to make it look like pine.
4. Give three reasons why a designer would use models.
5. Name three materials suitable for prototyping.

KEY WORDS

Make sure you understand these words before moving on!

- Two-dimensional (2D)
- Three-dimensional (3D)
- Isometric
- Oblique
- Exploded
- Orthographic projection
- Working drawings
- Dimensions
- Cross section
- Scale models
- Prototypes
- Virtual modelling

CAD Packages and CAM

The Internet and CD ROMs

The **Internet** is an excellent resource and search engines can make it easy to find the information you need. You should be able to describe how the Internet is used to...

- research the market place for design opportunities
- compare existing products
- conduct on-line surveys of potential customers
- check for availability of supplies
- market products using pop-ups, etc.
- sell products using websites.

CD ROMs are databases of information. Some CDs contain animations of industrial processes, video clips and comparative data. They can be useful sources of design information.

CAD Packages

Software such as ProDesktop, Pro Engineer, ArtCAM and 2D Design Tools allow you to draw directly onto a computer and create **virtual** 3D images. This is known as **computer aided design (CAD)**, and it also creates the numerical code needed to drive a range of **computer numerical control (CNC)** machinery.

CAD software is suitable for designers for many different reasons:
- Accurate drawings can be produced.

- Designs can be produced quickly by skilled operators.
- Information can be shared by email.
- Virtual reality images can be evaluated easily.
- The size or shape of a design can be changed quickly.
- Copies can be printed out for presentations to clients.
- Information can be easily stored.
- Copies can be backed-up for security.
- Designing can take place in different locations.
- The designs can be prototyped easily using CNC machines.

How CAM Works

Computer aided manufacture (CAM) uses numerical data called machine code. So, machinery used for CAM is often referred to as CNC, for example a CNC Lathe. Drawings are created using CAD packages, so the term CAD / CAM is often used to describe the process.

Three sets of data control the **X, Y** and **Z axis** of any CNC machine connected to the software. There are several advantages to this type of control:
- High speed production.
- High quality edges that need minimal surface finishing.
- Multiple items will be identical.
- Very accurate details can be cut.
- Changes to the programme can be made easily.

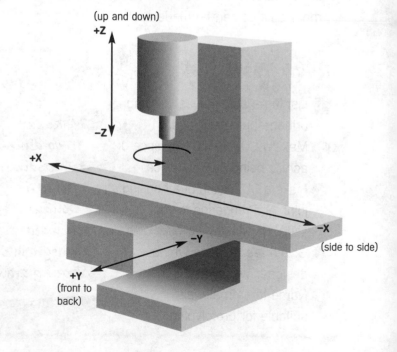

Two, Three and Four-axis Machines

Some machines use computer aided manufacture. A **two-axis** machine has two stepper motors that control the movements.

- The **X axis** controls the **sideways** movement.
- The **Y axis** controls the **front to back** movement.

Examples of two-axis machines commonly used in schools are lathes, engraving machines, plotters and vinyl cutters.

A **three-axis** machine also has a **Z axis** that controls the **up and down** movement. This means that more complex machining can take place. Routers and milling machines are three-axis machines commonly used in schools, mainly for sheet timber and plastics.

In a **four-axis** machine, the **fourth axis** allows the work to be **revolved** at the same time as it's being machined (it is very much like the addition of a lathe onto a milling machine). This means that full 3D can be achieved in one operation. These machines are available in some schools, and are commonly used in industry.

Two-axis Machine

Three-axis Machine

Four-axis Machine

Rapid Prototyping

Rapid prototyping or stereo lithography is a way of creating full 3D objects direct from a CAD drawing. A prototype is made by building up layers of wax.

They are used in industry, but not often in schools. You could create something similar using layers of paper cut on a vinyl cutter, but it's very time consuming.

Quick Test

1. What does CAD stand for?
2. Name one CNC machine.
3. Give four reasons why a designer may prefer to use CAD for designing rather than sketching.
4. Give four advantages of using CAM for manufacturing.
5. Sketch how a machine table moves in the X and Y axis.

KEY WORDS

Make sure you understand these words before moving on!

- Internet
- CD ROMs
- Virtual
- Computer aided design (CAD)
- Computer aided manufacture (CAM)
- Computer numerical control (CNC)
- X, Y and Z axis

Practice Questions

1 What kind of sketch would be most suitable for showing how to file a piece of acrylic? Tick the correct option.

A Orthographic ⬭

B Isometric ⬭

C Exploded ⬭

D Dimensioned ⬭

2 a) Complete this sketch of a drawing symbol.

b) What projection does it show? ....

3 Render the sketches below to show the material named.

a) Pine   **b)** Blue acrylic   **c)** Brass

4 Choose the correct words from the options given to complete the following sentences.

worker   orthographic   manufacturer   sketch   instructions   shapes   prototype

When a designer wishes to communicate with the ...., a(n)

.... drawing will be sent. This drawing will give the necessary

.... for a .... to be built.

5 Why is a working drawing more useful than a 3D sketch? Tick the correct options.

A Only a skilled draughtsman can understand the drawings ⬭

B It's easy to understand all the views ⬭

C They're very colourful ⬭

D Measurements can be taken from the drawings ⬭

E All the information for making is in the same place ⬭

F There's too much information in one place ⬭

6 Which of the following statements describe valid reasons for producing a model? Tick the correct options.

    A To help visualise the 3D appearance of the final product ◯

    B To see if it's the right scale ◯

    C To fill some spare time ◯

    D To consult clients and users ◯

    E To experiment with difficult materials and processes ◯

    F To avoid producing drawings and written work ◯

7 A cross-sectional view is shown of a fitting. Name the parts and complete the drawing.

8 Why would a designer choose to design and model their ideas on a computer? Tick the correct options.

    A It's slow to draw ◯

    B It's easy to share the information by email ◯

    C It creates virtual reality images for quick evaluation ◯

    D It's difficult to change the size or shape of models ◯

    E It's easy to print out copies for presentations to clients ◯

    F It's easy to draw this way ◯

    G There is the security of having back-up copies ◯

9 A CNC milling machine is sent data by a piece of software. Fill in the table to show which axis relates to each direction of movement.

Axis	Direction
a)	Left and right
b)	Front and back
c)	Up and down

Health and Safety

Signs

Signs have different shapes and colours that mean different things.

A **round red** signs means dangerous behaviour that you must avoid doing, e.g. don't smoke.

A **round blue** sign means a specific behaviour or action that must be carried out, e.g. must wear personal protective equipment.

A **triangular yellow** sign means be careful or take precautions, e.g. beware of dangerous chemicals or spills.

A **rectangular green** sign means escape routes, equipment and facilities, e.g. fire exit or first aid box.

COSHH

COSHH stands for the Control of Substances Harmful to Health.

Workshops can be very dangerous places to work in. You are responsible for your own health and safety, so you need to know all the different hazards there might be in a workshop, and how to protect yourself from them.

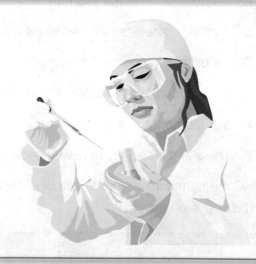

Safety Checks and Protection

Before you start to work on a machine you should check to see if there's any damage to the electric wiring or to the working parts.

Before you switch on the machinery, check that…
- any **machine guards** are in place
- you know where the power shut-off buttons are.

You should know when to wear protective clothing, such as…
- **goggles**
- **dust masks**
- **ear defenders**
- **latex gloves** or **gauntlets**
- **aprons** or overalls.

Hazards and Protecting Yourself

Your **eyes** can be at risk from chemicals, hot metal and dust. Safety options include wearing…
- safety spectacles
- goggles
- face-shields
- visors.

Your **ears** can be at risk from loud noises from machines. Safety options include wearing…
- ear defenders
- ear muffs.

Your **breathing** (lungs) can be at risk from dust, vapour and gas. All workshops should be fitted with dust extraction systems, which keep the air free from harmful dust. These systems are sometimes on the machines and must be switched on before working. Safety options include wearing…
- disposable filter dust masks
- a respirator
- air-fed helmets.

Your hands could be harmed by chemicals, hot metal, hot plastic or by carrying sharp / rough objects. To protect your hands you should wear…
- gauntlets
- rubber or latex gloves.

Your **body** could be harmed by chemicals, hot metal, spray from spray guns, knocking or cutting your body. Loose clothing could also get caught in a machine chuck. Safety options include…
- wearing an apron / leather apron / overalls
- rolling-up sleeves
- tucking in your tie and shirt, tying back your hair, etc.

Welding and Brazing

When heating metals (welding or brazing) you should protect yourself by wearing a leather apron, a face visor and gauntlets.

Never pick up a piece of metal from a heat treatment area with bare hands because even dark coloured metal, which appears cool, could actually be very hot.

Gluing Plastics

Solvents (e.g. Tensol) are often used to join plastics like polystyrene. The surface of the plastic is dissolved and then hardened by evaporation. This process gives off toxic vapour and is poisonous if it gets onto your hands.

To protect yourself, you should…
- wear a breathing mask and latex gloves
- always spread the adhesive with a spatula (never use your finger).

Health and Safety

Chemical Sprays

Paint and polish aerosols propel chemicals into the air using high pressure gas. The fine drops of paint or polish can be breathed in causing breathing problems.

To protect yourself, you should...
- wear a breathing mask and latex gloves
- use an extractor fan to remove the vapour
- always spray in a spray booth (the paint will go everywhere if it isn't contained).

Chemical Storage

All dangerous chemicals, for example many glues, polishes and paint brush cleaners, must be kept in a secure store cupboard. The cupboard should have...
- a sign on the door, e.g. caution, flammable or dangerous chemical
- a lock on the door.

CAUTION
Flammable material

CAUTION
Dangerous chemical

Working with Heat and Chemicals

If you're heating metals or plastics you should always wear body, eye and hand protection in order to prevent the heat from harming you.

If you're painting or polishing wood or joining plastics you should always wear eye, nose and hand protection to prevent the chemicals from harming you.

You should not open or tamper with machines that use high voltage.

You should know what these signs mean:

**DANGER
415 Volts**

Found on electrical machinery

Found on liquids

Eye protection must be worn

Wear dust mask

Hearing protection must be worn

Hand protection must be worn

Famous Designers

There are many designers, both past and present, who are inspirational to other designers, for example, the photographs show a street light by Charles Mackintosh and a lemon squeezer by Philippe Starck.

Street Light Designed by Charles Rennie Mackintosh

Lemon Squeezer Designed by Philippe Starck

You should know about the following designers:
- Charles Rennie Mackintosh
- Philippe Starck
- James Dyson
- Frank Lloyd Wright
- Ettore Sottsass.

In the Style of...

Retro styling is very popular. Designing in the style of past design movements is a great starting point, for example, the Hoover building is a famous piece of Art Deco architecture.

Hoover Building

You should know about the following styles:
- Bauhaus
- Art Deco
- Art Nouveau
- De Stijl
- Memphis
- Shaker.

Quick Test

1. What two things should you wear to protect yourself against dust?
2. What does this sign mean and where would you find it?
3. Why shouldn't you touch solvent cements with your bare hands?
4. What happens if you don't use a spray booth when using an aerosol?
5. What two precautions should you take before you switch a hand-sanding machine on?
6. Name three designers who have been inspirational in the last 100 years.

KEY WORDS

Make sure you understand these words before moving on!
- Machine guard
- Goggles
- Dust mask
- Ear defenders
- Latex gloves
- Gauntlets
- Apron

Social, Moral and Environmental Issues

Raw Materials

All raw materials, and the processes used to make products or materials, have an impact on human beings and the environment.

Many manufacturing processes involve heat that is generated by burning oil or coal. The gasses given off in this process can harm the atmosphere through **global warming**.

Harm can also occur when raw materials are mined, felled, extracted from oil wells or grown as crops.

Designers have a responsibility to make sure that ideally no person, animal or part of the planet is harmed by the product that they design.

Finite Resources

Some raw materials can't be replaced. They are known as **finite resources** and include...

- coal – used to generate electricity and smelt metal ores
- oil – used to generate electricity and as the base for many plastics
- metal ores – the raw materials are made into metals.

To avoid using up all our finite resources we should always consider...

- recycling materials
- re-using products that can be taken apart and used again (termed **disassembly**).

Ethical Employers

The living and working conditions of the people who extract and work raw materials are also important. Many companies are **ethical employers** who ensure that their workers get fair wages and safe working conditions.

Manufacturing can be a noisy and dirty process that can cause **industrial injuries,** for example...

- deafness
- lung problems.

Providing people with work is beneficial as it allows them to earn money and support themselves. But, a **moral dilemma** is created if the manufacture of the products causes harm to people or animals.

Carbon Footprint

Many manufacturing activities generate carbon by burning fossil fuels. This is called their **carbon footprint.** The carbon needs to be offset through schemes like planting trees or **carbon capture.** Companies in the future will be expected to be **carbon neutral.**

Social, Moral and Environmental Issues

The 6 Rs

In order to minimise the environmental impact of using raw materials, designers should consider the '6 Rs':

- **Reduce** the amount of material used in manufacture.
- **Recycle** the materials already used by melting or re-processing.
- **Re-use** by designing for disassembly and recovering materials from 'end of life' products.
- **Repair** products rather than replacing them.
- **Refuse** to accept unethical or wasteful designs.
- **Re-think** our attitude to environmental impact.

Maintenance

Designers should build products that can be repaired and **maintained** rather than opting for a 'throw-away' attitude known as **planned obsolescence**.

Sustainability and Eco Design

Resources are finite so designers have to take account of **sustainability**. If we are careful the finite resources will last longer and alternatives may be found.

Houses and manufacturing systems are now being developed which…

- are carbon neutral
- have a minimal impact on the environment.

Quick Test

1. State one thing that an employer could do to become ethical.
2. Why is producing carbon an issue for a designer?
3. Suggest three things that a designer could do to make their design more environmentally friendly.
4. What is the difference between re-using a material and recycling it?

KEY WORDS

Make sure you understand these words before moving on!

- Global warming
- Finite resource
- Ethical employers
- Industrial injury
- Moral dilemma
- Carbon footprint
- Carbon capture
- Carbon neutral
- Reduce
- Recycle
- Re-use
- Repair
- Refuse
- Re-think
- Maintained
- Planned obsolescence
- Sustainability

Cultural and Moral Issues

Human Factors

Most products are designed for humans, so when designing a product you would need to consider human factors, for example…

- physical ability
- health
- intellectual ability.

If you're designing entirely for animals then a similar investigation might be needed, but you may find relevant information harder to find.

Inclusive and Exclusive Design

An **ideal product** is one that can be used by everyone. This is known as **inclusive design** and, although it's an impossible aim, designers should design products that are suitable for as many people as possible.

Some products have to be designed to take into account the **specific** needs of people, e.g. the very young or the elderly. Examples of **exclusive design** are car seats for babies and wheelchairs for disabled people.

Stereotypes

Everybody is unique (different), but people are often **stereotyped** (classed in groups). These stereotypes can be useful to designers, manufacturers and especially retailers.

Ensuring that products are aimed at particular market groups is common practice.

Offence

Designers have a responsibility to make sure that their designs will not offend the people who use the products. Extensive research is carried out to find out what cultural issues are involved before products are marketed.

Ethical Trading

Retailers who consider the welfare of the planet and the people who live on it are considered to be working towards a **sustainable** future.

By looking after the planet and not wasting valuable resources, it's more likely that…

- a greater number of people will be able to survive
- everyone will have a better quality of life.

Anthropometric Data

Anthropometrics

Anthropometrics is the study of human measurements. Measurements have been taken from millions of people of different shapes and sizes and put together in charts. Designers try to cater for 90% of the population.

An **ergonome** is an adjustable scale template that can be drawn round. It is used to help a designer to work to scale using the average human measurements.

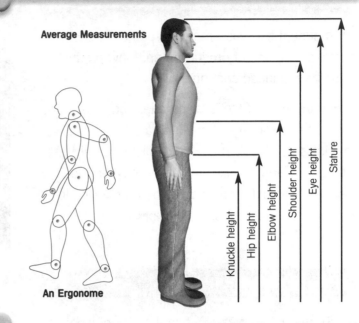

Average Measurements

Knuckle height
Hip height
Elbow height
Shoulder height
Eye height
Stature

An Ergonome

Ergonomics

Ergonomics is the study of the efficiency of people in their working environment. It often deals with the application of anthropometric data.

Ergonomics deals with issues such as comfort and safety, for example…
- what colour is best for safety equipment?
- how much weight can one person safely lift?

The Design Cast List

Clients commission a piece of work. You would carry out an interview with the client to find out exactly what they need from a product.

A client could…
- represent a large company wishing to introduce a new range of furniture
- be an individual, e.g. one of your parents who wants you to make a storage system.

The **designer** will then come up with ideas for the new product and share them with the client. Once an idea has been agreed with the client, the designer solves any design / technical problems and passes the details on to the **manufacturer.**

Different manufacturers specialise in certain types of production. A designer may need to work with more than one manufacturer in order to have the product made.

Anyone who buys or has access to the product becomes a **user** of the product. Many products have worldwide distribution, but some may only be used by a few people at home. It's important that as many people as possible are satisfied with the product. **User surveys** are carried out frequently to see if improvements could be made during the lifetime of a product.

Scales of Production

'One-off' Production

It's important that you are aware of the various possible methods of production and how products could be produced commercially.

A **'one-off' production** is when one product is made at one particular time. It could be a prototype or a very intricate object. These productions usually take a long time and so often result in an expensive product. A typical product could be a prototype to test the market or a piece of sculpture.

Batch Production

Batch production occurs when a series of identical products are made together in either small or large quantities. Once made, another series of products may be produced using the same equipment and workforce. A typical example is a piece of furniture, e.g. a stool.

Mass Production

Mass production involves the product going through various stages on a production line where the workers at a particular stage are responsible for a certain part of the product. The product is usually produced in large numbers for days or even weeks. The product will be less expensive because of the large scale of production. A typical example is a car.

Continuous Production

Continuous production is where the product is continually produced over a period of hours, days or even years ('24/7'). The product will be relatively inexpensive.

Typical examples are supermarket processed foods and wood screws.

'Just-in-time' Production

'Just-in-time' production involves the arrival of component parts at exactly the time they're needed at the factory. 'Just in time' uses less storage space so saves on costly warehousing. But, if the supply of components is stopped, the production line is interrupted, which then becomes very costly.

Quality Assurance and Quality Control

Quality assurance (QA) checks the production systems before, during and after manufacture. It ensures that consistency is achieved and that a product meets the required standards.

The customer is an important part of any QA system and may be involved in the monitoring at various stages (e.g. through user surveys and focus groups).

Quality control (QC) guarantees the accuracy of a product by using jigs, moulds, etc and by a series of checks carried out on a product as it's made. The checks make sure that each product meets a specific standard and could include…

- dimensional accuracy
- taste (food)
- material quality
- electrical safety / continuity
- flammability tests.

Testing is an important part of the manufacture of a product and can take place at any time during production (**sampling**). For example, an injection-moulded plastic bottle top could be tested after ten, a thousand or a million have been produced. Some of the tests would include checking its diameter, thickness and whether it screws onto its container properly.

Tolerances

When products are produced in large quantities, it's very hard to guarantee that each one will meet the specifications accurately. A **tolerance** has to be accepted which specifies the minimum and maximum measurements.

The analysis of tolerance tests can…

- signal the imminent failure of a machine or a tool
- help to achieve the ultimate aim of quality control, which is **zero faults**.

Quick Test

1. What do the letters QA stand for?
2. State two ways in which members of the public can contribute to QA.
3. What do the letters QC stand for?
4. State two ways in which factory tests can contribute to QC.
5. Name the best production method for making tins of dog food.

KEY WORDS

Make sure you understand these words before moving on!

- Inclusive design
- Exclusive design
- Stereotypes
- Anthropometrics
- Ergonomics
- Client
- Designer
- Manufacturer
- User
- 'One-off' production
- Batch production
- Mass production
- Continuous production
- 'Just-in-time'
- QA
- QC
- Sampling
- Tolerance

Quality Assurance Systems

Accuracy

Holding and guiding devices are used when manufacturing in quantity. They're often used for…
- drilling holes accurately in components
- cutting materials to size.

You may have made a jig or a fixture to enable you to save time and to ensure accuracy when making your controlled assessment.

Jigs

Jigs are often commercially made for a variety of similar jobs. They are used for guiding a tool.

Jigs can be…
- specially made for a component
- made adjustable for a range of similar operations.

A **mitre box** is used for cutting angles accurately when making mitred joints, for example picture frames.

A **sawing jig** is an adjustable jig used for sawing accurately at a variety of angles.

Mitre Box

Sawing Jig

Fixtures

Fixtures are similar to jigs, but these holding devices are fixed to machines to aid quantity production.

A sawing fence is used on a band saw to guide the material in a straight line at a given width.

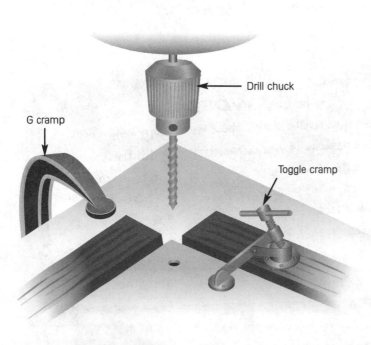

Drill chuck

G cramp

Toggle cramp

Flow Symbols and Charts

The following symbols show the three most common parts of the manufacturing process.

Start/End

Process

Decision

When planning the manufacture of a product, the designer needs to show which processes happen at what time.

A **flow chart** may…

- makes things clearer
- help prevent mistakes, e.g. polish being applied before the joints have been cut.

Flow Chart Showing Material Being Polished

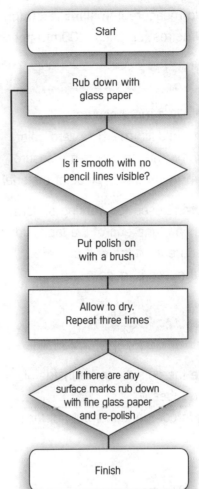

Start

Rub down with glass paper

Is it smooth with no pencil lines visible?

Feedback loop to correct any mistakes before going on to the next process. The criteria should be included for accurate checking.

Put polish on with a brush

Allow to dry. Repeat three times

If there are any surface marks rub down with fine glass paper and re-polish

Finish

Gantt Charts

Gantt charts are block charts that are useful if more than one process is being completed at the same time, so every sub-component is finished ready for the final assembly.

Process	Week 1	Week 2	Week 3	Week 4
Marking out	■			
Cutting		■		
Joining			■	
Finishing				■

Measuring and Checking

Accuracy

It is essential that all of your measurements are **accurate**. You should measure in **millimetres** (mm) as opposed to centimetres (cm), e.g. 200mm not 20cm. Always check measurements, especially where parts need to fit together.

Steel Rule and Tape Measure

A **steel rule** is generally a more accurate measuring tool than a plastic rule. The zero point begins at the end of the rule so you can measure more accurately from the end of a piece of material.

A **steel tape measure** is used for working on larger sized materials. The lip at the end of the tape slides to allow you to measure from the end of the material or against a raised surface.

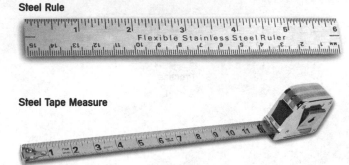

Steel Rule

Steel Tape Measure

Calipers, Micrometer and Vernier Gauge

Calipers are...
- used to measure the outside or the inside of circular bars and tubes
- suitable for simple comparisons, but not for precision tasks.

A **micrometer**...
- is accurate up to 0.01mm
- can usually only measure outside dimensions
- has a ratchet mechanism, which prevents the jaws being pressed too hard
- is available only for a given range, e.g. 0 to 50mm and 50mm to 100mm.

A **Vernier gauge** is...
- very accurate for measuring both inside and outside dimensions
- available in digital versions.

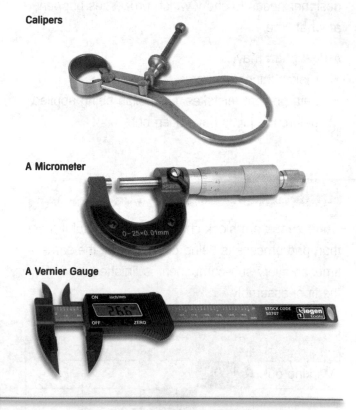

Calipers

A Micrometer

A Vernier Gauge

Spirit Level

A **spirit level** is a tool used for checking horizontal and vertical surfaces. For example, they are especially useful when fitting shelves. The bubble in the tube must fall within the markings.

Measuring

If a lot of identical items are manufactured, you will need to undertake some form of measuring check to make sure that each part is identical. This might be one part of your quality assurance procedures.

Steel rules and calipers can be used to check measurements, but making a special tool is often easier if lots of identical measurements are needed.

Measuring Stick

A simple cut length of material can be used to check lengths. Two sticks (cut with pointed ends) can be used for checking that the diagonal distances on a box are the same. This shows that the box is square and can be more accurate than using a tri square.

Gap Gauge

A **gap gauge** is another simple checking tool that can be used to make sure components are the correct size. If the piece is a tight fit in the gap, it is the correct length.

Industrial Measuring Devices

Many manufacturing industries need to use very sophisticated technologies to measure and check dimensions.

For example...
*   laser measuring devices
*   ultrasonic measuring.

Quick Test

1. What device would you use to measure the outside diameter of a tube $\pm\frac{1}{2}$mm?
2. What device would you use to make sure a picture was hanging level?
3. What is the most accurate measuring device available to you?
4. What is the purpose of a gap gauge?

KEY WORDS

Make sure you understand these words before moving on!
*   Jig
*   Fixture
*   Flow chart
*   Gantt chart
*   Caliper
*   Micrometer
*   Vernier gauge
*   Spirit level
*   Gap gauge

Practice Questions

1 For each of the following symbols, explain what it means and why it is important in terms of safety.

	a)	
	b)	
	c)	

2 Circle the correct options in the following sentences.

When working in a hot metal area it's essential that **goggles** / **gauntlets** are worn to protect the hands, and a visor is used for the **face** / **body**. The **dust** / **water** extraction system should be switched on and any **hot** / **cold** metal should be placed **in a safe area** / **on a desk** so that no one picks it up by mistake.

3 Give two reasons why you might look at the work of modern designers.

a) ....

b) ....

4 The table contains the names of six environmental issues considered by designers. Match descriptions **A, B, C, D, E** and **F** with the environmental factors **1–6** in the table. Enter the appropriate number in the boxes provided.

Environmental Factors			
1	Reduce	**4**	Repair
2	Recycle	**5**	Refuse
3	Re-use	**6**	Re-think

A Looking at things from a different point of view and not taking things for granted

B Not throwing things away because they're broken

C Encouraging users to dispose of unwanted things responsibly

D Using fewer raw materials and cutting down on energy costs

E Building things that can easily be taken apart and separated into different materials for disposal

F Not accepting products that haven't been produced in an environmentally friendly way

5 Why would a designer choose to design ethically? Tick the correct options.

A It cuts down on waste ⬜

B They can produce products more cheaply ⬜

C It makes them more popular with their friends ⬜

D It will use up natural resources more quickly ⬜

E The health and safety of the workers is important ⬜

F It will reduce their carbon footprint ⬜

6 Complete the 'design cast list' table below:

Person	What They Do
Designer	a)
Client	b)
Manufacturer	c)
User	d)

7 Which systems do manufacturers use to ensure that every item produced is of the same high quality? Tick the correct options.

A Quality dimensioning ⬜   B Quality assurance ⬜

C Quality control ⬜   D Quality inspection ⬜

8 Explain when batch production is used.

9 Give two reasons why it would be useful to draw a flow chart to show the process of finishing a piece of wood.

a)   b)

Holding Devices

Vices

A **vice** is used to hold material in place while you are working on it. This is very important in terms of efficiency and safety.

A **woodworking vice** has wooden jaws and is used to hold timber and plastics to the workbench while they're being cut and shaped.

A **metalwork** or **engineering vice** is raised above the workbench and is available with…
- extra-hard steel jaws for heavy metalworking
- fibre (soft) jaws for lightweight metal, e.g. brass, aluminium and especially plastics.

A **machine vice** is used for holding materials while they're being drilled or milled.

A **hand vice** is used to hold smaller and irregular pieces of plastic and sheet metal that will not fit into a machine vice.

Woodworking Vice

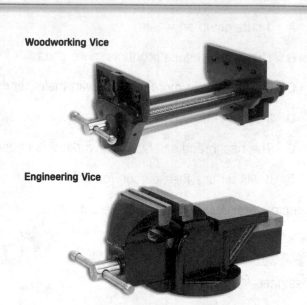

Engineering Vice

Machine Vice

Cramps

A **G cramp** is used for holding materials onto bench tops while working or gluing.

A **sash cramp** is used for holding wooden joints together while gluing.

A **speed cramp** is similar to a sash cramp, but it uses a self-locking system which makes the positioning easier.

A **corner cramp** is used for holding materials at right angles whilst joining them together.

A **toggle cramp** is found on vacuum forming machines. It's useful…
- when making jigs for volume production
- for cramping small pieces of material on a drilling machine.

G Cramp

Sash Cramp

Speed Cramp

Corner Cramp

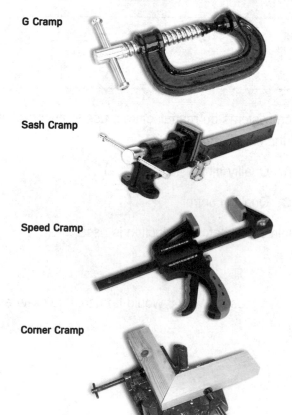

Hammers

Hammers are designed for a variety of jobs.

A **claw hammer** is a heavy hammer, which can drive large nails through timber. The claw is used to remove bent nails.

A **cross pein hammer**...
- is a general purpose hammer that gets its name from its wedge-shaped rear face
- is used to start small nails and pins which are held between the fingers.

A **ball pein hammer**...
- has a rounded pein as the rear face
- is used for spreading or rounding rivet heads and other pieces of metal
- is available in very large sizes for heavy duty work, e.g. hot metalworking.

A **planishing hammer**...
- has a very polished head so shouldn't be used for general hammering
- is used with a polished stake for finishing beaten metalwork.

Mallets

Mallets are usually made entirely from beech or boxwood. They are used...
- for driving chisels and gouges without damaging the handle
- when assembling wooden joints.

A **rubber / nylon mallet**...
- is the modern equivalent of the traditional beech mallet
- can be used for assembling wood joints
- can be used to bend over sheet metals without damaging the surface.

A **bossing mallet**...
- is a wooden mallet made with an egg-shaped boxwood head
- is used with a leather sandbag for hollowing, or dishing sheet metal.

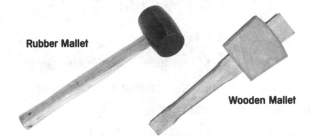

Rubber Mallet

Wooden Mallet

Quick Test

1. Why would you never hit a woodworking chisel with a hammer?
2. What is the ball pein on a hammer used for?
3. What needs to be done to a metalworking vice before it can be used to hold a piece of brass?
4. Sketch and name the cramp that you would use for holding the corner of a picture frame while the glue sets.

KEY WORDS

Make sure you understand these words before moving on!
- Woodworking vice
- Metalwork or engineering vice
- G cramp
- Hammer
- Mallet

Properties and Environmental Issues

Choosing a Material

There are many properties that you should consider when choosing a material for a particular task:

Strength – withstanding force without breaking or bending permanently.

Tensile strength – withstanding force when stretched.

Compressive strength – withstanding force when being crushed.

Durability – withstanding wear and tear and weathering.

Flexibility – how easily a material will bend or distort.

Elasticity – the ability to regain its original shape after it has been deformed.

Plasticity – changing in shape permanently without cracking or breaking.

Malleability – the ability to be easily pressed, spread and hammered into shapes.

Ductility – the ability to be stretched and permanently deformed without breaking.

Brittleness – how easily a material will break without bending (the opposite of ductile).

Hardness – resistance to scratching, cutting, denting and wear.

Work hardness – when the structure of the metal changes as a result of repeated hammering or strain.

Toughness – resistance to sudden shock without breaking or deforming.

Impact resistance – resisting denting.

Shear – strong sliding forces acting opposite to each other.

Stress – any forces acting on a material.

Electrical conductivity – how easily a material allows electricity to flow through it.

Thermal conductivity – how easily a material allows heat to flow through it.

Chemical resistance – resists chemical attack.

Wood

Wood is a renewable resource that can be replaced by replanting trees. Usually two saplings are planted ('**two for one**') to replace each tree cut down, and the weaker one pulled up later. Natural rainforest is a precious resource and shouldn't be destroyed.

Trees store carbon in their trunks (carbon capture), and if the wood rots or burns this carbon is released into the atmosphere. So, rather than being burned, wood products should be re-used or recycled.

Wood dust, chips or off-cuts can be recycled into chip board or Medium Density Fibreboard (MDF).

Properties and Environmental Issues

Metals

Metals come from non-renewable ores that are mined from the ground. There are health and safety issues with the working conditions of the miners digging in dangerous places and breathing in dust.

Ores need to be smelted to provide the pure metal. The heat energy required to smelt and shape metals usually comes from burning non-renewable **fossil fuels**, which give off toxic fumes and release carbon dioxide into the atmosphere. The smelting process also gives off toxic fumes.

So, recycling metals is important in order to...

* preserve the limited resources
* reduce the amount of energy needed to work the metals.

For example, Bauxite (aluminium ore) is a plentiful metal, but producing new aluminium from bauxite uses nine times as much energy as melting recycled cans.

Plastics

Most plastics are made from oil (a non-renewable resource). Stocks of oil are now running out so alternative sources have been found in plant material, known as **bio fuel**. But, growing plants to provide bio fuel or starch polymers, uses land that could have been planted for food. This increases food prices and can lead to food shortages.

Energy is used to work and produce plastics. The **toxic by-products** also have to be disposed of.

Decisions have to be made about whether plastics are the best material to choose for making products. Most plastics can be recycled, but because there are so many different types it's difficult to separate them for the different processes required.

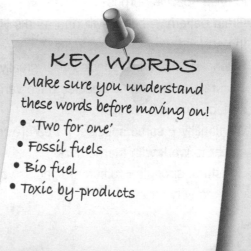

Quick Test

1 Why are material properties important to a designer?
2 State three material properties of a piece of steel.
3 Why shouldn't wood from a tropical rainforest be used for making a piece of furniture?
4 Why should we encourage recycling of aluminium cans?
5 What will happen to the environment if we continue to produce and use oil-based plastics?

KEY WORDS
Make sure you understand these words before moving on!
* 'Two for one'
* Fossil fuels
* Bio fuel
* Toxic by-products

Woods

Timber

Timber is the general name for wood materials. There are three main types of timber.

Hardwoods come from **deciduous** or broad-leafed trees. They are generally slow growing, which tends to make them harder. But, some hardwoods can be light and soft, for example Balsawood.

Softwoods come from **coniferous** trees, which have needles rather than leaves. Softwoods generally grow faster than hardwoods and are usually softer to work.

Manufactured boards are timber sheets that are made either by gluing together wood layers (**veneers**) or wood **fibres**.

Manufactured boards have been developed mainly for industrial production techniques because they can be made in very large sheets of consistent quality.

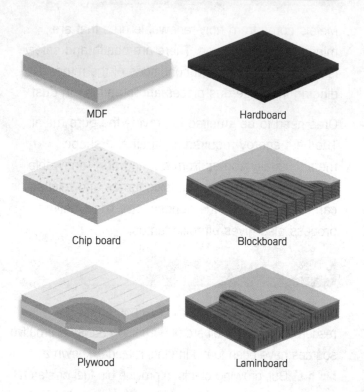

MDF

Hardboard

Chip board

Blockboard

Plywood

Laminboard

Natural Characteristics

Wood grows with fibres that run along the length of a tree and grow in groups, which make up annual rings.

The wood will split easily **along the length** of the fibres, but it must be cut or sawn **across the grain**.

Here are some things to consider when choosing wood for a specific purpose:
- Grain pattern – the growth ring marks visible on the surface.
- Colour – different tree species differ greatly in colour.
- Texture – different tree species have varied surface and cell textures.
- Workability – some species of tree are much easier to work with than others.
- Structural strength – different species vary from weak to very strong.

Hardwoods

Name	Description	Uses
Beech	• A straight-grained hardwood with a fine texture • Light in colour • Very hard but easy to work with • Can be steam bent (i.e. heated in steam and shaped by cramping around a former until cool)	• Furniture • Toys • Tool handles
Oak	• A very strong, light-brown wood • Open grained • Very hard, but quite easy to work with	• High quality furniture • Beams used in buildings • Veneers
Ash	• Open grained • Easy to work with • Pale cream colour, often stained black • Can be laminated (i.e. sliced into veneers which are glued together and cramped around a former until dry)	• Tool handles • Sports equipment • Furniture • Ladders • Veneers
Mahogany	• Reddish-brown in colour • Easy to work with	• Indoor furniture • Shop fittings • Bars • Veneers
Teak	• A very durable, oily wood • Golden brown in colour • Highly resistant to moisture	• Outdoor furniture • Boat building • Laboratory furniture and equipment

Woods

Softwoods

Name	Description	Uses
Scots pine (Red Deal)	• Straight-grained but knotty • Light in colour (cream / pale brown) • Fairly strong but easy to work with • Inexpensive	• Readily available for DIY work • Mainly used for constructional work and simple joinery
Parana pine	• Hard and straight-grained • Almost knot free • Fairly strong and durable • Expensive • Pale yellow with red / brown streaks	• High quality pine furniture and fittings, e.g. doors and staircases
Western red cedar	• Light in weight and knot free • Reddish brown • Easy to work with, but weak and expensive • Naturally oily	• Outdoor uses, e.g. timber cladding of buildings, fencing
Pitch pine	• Pale-yellow coloured with dark lines and a fine, even texture. • Medium in weight • Stiff and stable	• Furniture • Church pews • Veneers
Yew	• Rich red-brown with small hard knots • Very durable	• High quality furniture used in bedrooms and kitchens

Manufactured Boards

All manufactured boards are made from real wood that has been processed. Some use waste materials (**chip board** or **MDF**) whilst others are cut into veneers and processed to make the boards (**plywood**). Other boards are **hardboard** and **blockboard**.

Advantages of using manufactured boards:

- Available in large sizes.
- Stable (they don't warp or twist).
- Easy to work with hand and machine tools (though edges need extra care).
- Can be painted or polished to give a high quality finish.

Disadvantages of using manufactured boards:

- Can be more expensive than solid wood.
- Not as attractive as solid wood.
- The edges need to be covered to hide the inside.

Chip Board

MDF

Name	Description	Uses
Medium density fibreboard (MDF)	• Has a smooth, even surface • Easily machined and painted or stained • Available in water and fire-resistant form • Often veneered or painted to improve its appearance	• Furniture and interior panelling
Hardboard	• A very cheap particle board • Can have a **laminated** plastic surface	• Kitchen unit back panels
Chip board	• Made from chips of wood glued together with urea formaldehyde • Usually veneered with an attractive hardwood or covered in plastic laminate	• Kitchen and bedroom furniture • Shelving and general DIY work
Plywood	• A very strong board, constructed of layers of veneer or plies, which are glued with the grains at 90° to each other • Interior and exterior grades available • A very durable water and boil proof (WBP) plywood can be used in extreme conditions	• Furniture making • Boat building and exterior work
Blockboard	• Similar to plywood, but has a central layer made from strips of timber • Used where heavier structures are needed	• Shelving and worktops • Furniture backs

Marking Out Wood

Measuring Lengths and Angles

A **steel rule** is used to...
- measure lengths when marking on wood
- set gauges and a compass.

A **tri square** is used...
- to mark lines at exactly 90° from the edge of the material
- to check that the material has square corners
- as a checking tool when assembling components that need to be square.

Before you can use a tri square you need to have one accurate straight edge on the material. It's also available as a mitre square, pre-set at 45°.

A **sliding bevel** is an adjustable angle marker that can be set to any angle.

Card **templates** are used for marking out curved shapes onto any material. You can make symmetrical shapes using the following method:
1. Fold the template in half and cut both sides together.
2. Unfold and draw round the curve.

Steel Rule

Tri Square

Sliding Bevel

Card Template

Marking the Surface

A pencil is used for marking wood. It's soft and will not mark the surface too deeply.

A **marking knife** is a sharp knife used to cut the wood fibres before cutting across the grain.

The cut line...
- stops the wood tearing when it is sawn
- is useful as a ledge to put cutting tools in for accuracy.

Gauges

A **marking gauge** is a simple tool for scribing lines parallel to a straight edge on timber. One version has a drilled hole for a pencil so it can be used to mark more visible lines.

A **cutting gauge**...
- is a simple tool for scribing lines across the grain on timber
- cuts the fibres so they will not tear when sawn.

A **mortise gauge** is a tool that scribes two parallel lines at once for marking out both sides of a joint.

Rosewood Mortise Gauge

Chiselling Wood

Chisels need a sharp edge to slice across the grain. There are four types of wood chisel:

- A **firmer chisel** is a general hand tool.
- A **bevel-edged chisel** is used for corners that are less than 90°, e.g. a dovetail joint.
- A **mortise chisel** is a strong, thicker chisel used for chopping deep holes for joints.
- A **gouge** has a curved blade for carving (can be sharpened on the inside or the outside).

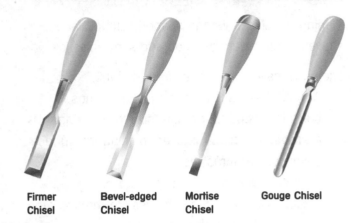

Firmer Chisel   Bevel-edged Chisel   Mortise Chisel   Gouge Chisel

Basic Chiselling Actions

There are several basic chiselling actions:

- **Horizontal paring** is cutting across a joint to clean out waste.

- **Vertical paring** is pushing down onto a waste surface to shape the end of a piece of wood.

- **Chopping** is digging out waste from a mortise by cutting the fibres into short lengths.

Mortising machines can be used to cut deep recesses for joints.

Hitting a chisel with the steel face of a hammer will damage it, so mallets made from beech or boxwood are used.

Quick Test

1. Sketch a 3D view of a piece of 3 plywood.
2. What does the acronym MDF stand for?
3. What are MDF and hardboard made from?
4. Which three tools are used to mark a line around the end of a piece of wood?
5. Why does a woodworker use a marking knife?

KEY WORDS

Make sure you understand these words before moving on!

- Veneers
- Fibres
- Annual rings
- Chip board
- MDF
- Plywood
- Hardboard
- Blockboard
- Laminated
- Steel rule
- Tri square
- Sliding bevel
- Templates
- Marking knife
- Marking gauge
- Cutting gauge
- Mortise gauge
- Firmer chisel
- Bevel-edged chisel
- Mortise chisel
- Gouge
- Horizontal paring
- Vertical paring
- Chopping

Cutting Wood

Planing

Planing is using a wedge-shaped cutting blade to shave off thin layers of wood and some plastics.

There are several varieties of hand plane.

- A **jack plane** is used to remove shavings to bring the wood to the right size for working.
- A **smoothing plane** is used to finish the surface before using abrasives.
- A **block plane** is used for planing end grain.
- A **rasp and surform** can be used to remove large amounts of wood when sculpting or carving.

There are some specially adapted planes designed for specific tasks:

- A **shoulder plane** is used to clean up a rebate.
- A **spokeshave** is used for curved surfaces.

Schools may use a powered planing machine, which works by using a rotary cutter. There are also hand-held versions available.

Powered Planing Machine

Smoothing Plane

Block Plane

Spokeshave

Drilling

A drilling machine, hand drill or power drill makes a hole by rotating a drill or boring bit into a material. It's important to match the correct drill bit to the material. Drill bits (cutters) are usually made from carbon steel or high speed steel (HSS).

The drill bit is rotated in a clockwise direction either by hand or by an electrically-powered device and is pressed onto the material surface. Drill bits are designed to cut and remove the waste material.

Battery-operated Drills	Mains-operated Drills
Positive Points	**Positive Points**
• No training lead	• Will go on forever
• Can use outside	• More powerful
• Non lethal	
Negative Points	**Negative Points**
• Has to be re-charged	• Limited to supply area
• Has less reserve of power	• Trip hazard
	• Not as safe in damp weather
	• Lethal

Centre bits and Jennings type auger bits are used with a carpenter's brace.

Forstner bits can be used on timber and some plastics to produce clean, flat-bottomed holes.

Twist drills or **jobbers drills** are used for drilling smaller diameter holes in timber. They are unsuitable for larger diameters in wood as they leave a ragged edge to the hole.

Countersink bits are used to allow countersunk screw heads or rivets to finish flush with the surface of the material.

Hole saws are used for cutting large diameters in thin materials.

Power Drill

Drilling (Cont.)

Pedestal drills (also known as **pillar drills** and drill presses) can be bench or floor mounted.

They provide the safest and easiest method of drilling materials that can be lifted onto the drilling table. During drilling the material must be firmly held in place by using a vice or cramp.

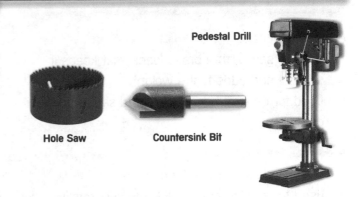

Pedestal Drill

Hole Saw   **Countersink Bit**

Sawing Wood

Saws have triangular-shaped teeth so they remove a small amount of material on the forward stroke. The teeth are '**set**' to make a cut wider than the blade to reduce friction.

Ideally, at least three teeth should be on the material at any time. Wood cutting saws often have the handle fixed directly to the blade.

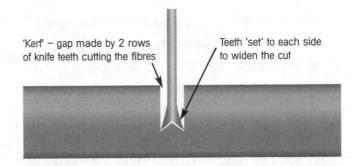

'Kerf' – gap made by 2 rows of knife teeth cutting the fibres

Teeth 'set' to each side to widen the cut

Hand Saws

A **rip saw**...
- has teeth-like chisels that cut **along** the grain
- is used for sawing large pieces of wood to width.

A **cross cut saw**...
- has teeth-like knives that cut **across** the grain
- is used for sawing large pieces of wood to length.

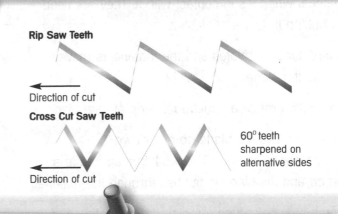

Rip Saw Teeth

Direction of cut

Cross Cut Saw Teeth

Direction of cut

60° teeth sharpened on alternative sides

Quick Test

1. When would you use a jack plane?
2. Sketch a cross-cut blade.
3. What tool would you use to make a 50mm hole in a piece of pine?
4. What is the difference between sawing along the grain of a piece of wood and sawing across it?

KEY WORDS

Make sure you understand these words before moving on!
- Jack plane
- Smoothing plane
- Block plane
- Rasp and surform
- Shoulder plane
- Spokeshave
- Centre bits
- Forstner bits
- Twist drills
- Countersink
- Hole saws
- Pillar drills
- Rip saw
- Cross cut saw

Cutting and Joining Wood

Back Saws

A **tenon saw** is…
- a small saw with a brass 'back' that keeps it straight and adds to the weight
- used for general bench work and sawing out joints.

A **dovetail saw** is…
- a smaller version of a tenon saw
- used for sawing out small joints accurately.

A **coping saw**…
- has a slot-in blade (the blade is held in tension within a frame) that can be easily changed when worn or damaged
- is used for cutting round curves
- works on the backward stroke, cutting when pulled.

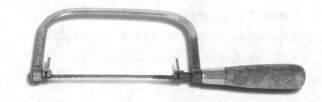

A Tenon Saw

A Coping Saw

Power Saws

A power saw has a fast-moving blade that cuts the material when it's in contact with it. They are used for cutting timber and plastics.

A **circular saw** rotates and the material is moved across the blade.

Bandsaws rotate a continuous strip of saw blade.

Jigsaws move the blade up and down (reciprocating motion). The work is cramped to a bench and the blade is pushed through the material.

Scroll saws also use a reciprocating motion, but the blade is held in tension and moves up and down through a table that can be angled. The wood is pushed past the blade.

Bandsaw

Jigsaw

Scroll Saw

Joints

Many traditional joints are used to build structural strength into products.

The strength of a joint increases with…
- surface area
- the use of glue.

So, stronger joints can be complicated.

Some traditional wood joints are being replaced with new methods, which are quicker to produce and are often stronger. This trend supports an increase in flat-pack products.

Simple Joints

A **butt joint**…

- is simple, but weak
- can be mitred (cut at 45°)
- is often used in picture frames.

A **halving joint**…

- comes in several variations
- is used to make a frame
- has half the material removed from each piece using a saw and chisel
- can be strengthened with a dowel through the joint.

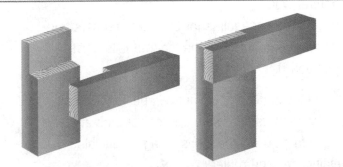

A **dowel joint**…

- is very easy to produce
- uses aligned holes and pegs (dowels). Some commercial products use serrated plastic dowels for home assembly.
- is used in frame and carcase joints for chair legs and cupboard corners
- is drilled using a jig to ensure accuracy.

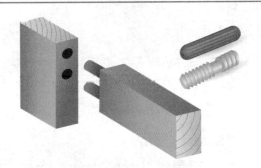

A **lap joint** is…

- stronger than a butt joint (has a bigger surface area for gluing)
- often strengthened with nails
- used in frame and carcase joints.

A **housing joint**…

- has a simple slot cut into one piece (to make a support for a cupboard shelf and to increase the gluing area)
- is often made with an electrically-powered router.

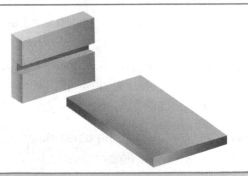

Joining Wood

Complicated Joints

The **mortise and tenon** joint is...
- a strong joint
- used in frame joints for chair and table legs.

In commercial production, the mortise is milled out so the tenon is machined with a rounded edge.

The **dovetail joint** is...
- the strongest joint for box constructions in natural wood
- used in carcase joints for cupboard corners and drawer constructions (the tapered tails don't pull apart).

This joint looks decorative, but can be very difficult to cut by hand using a saw and chisel. Jigs are available to help cut the joints using a special dovetail cutter and router.

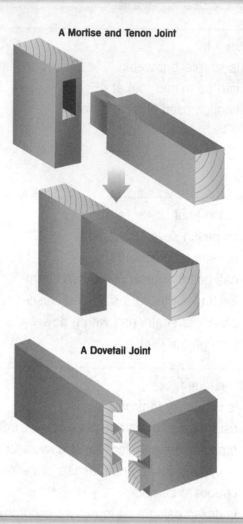

A Mortise and Tenon Joint

A Dovetail Joint

Temporary Joints

There are many different types of **nail**, but they make very weak joints if used on their own.

Nails or nailed joints are used for...
- holding wood together while the glue dries
- fixing the backs of cupboards
- decorative mouldings
- general building and DIY work.

A **screw** can be very strong when used across the grain. There are different types, but cross head screws are increasingly used instead of slot heads because they're easier to drive in by hand or by using an electrically powered driver.

Screws are useful for fixing other materials, such as metals or plastics, to timber.

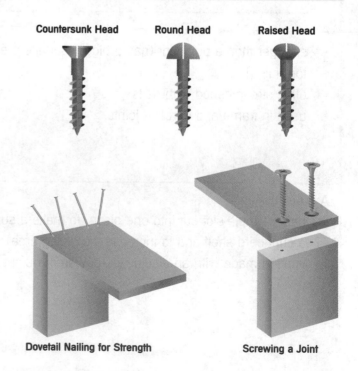

Countersunk Head　　**Round Head**　　**Raised Head**

Dovetail Nailing for Strength　　**Screwing a Joint**

Permanent Joints

A **biscuit joint** is a quick and easy method of joining boards…

- at right angles
- side by side to make wider boards.

Slots are made in the board using an electrically powered cutter. The 'biscuits' are elliptical pieces of timber, which have been dried and compressed then fixed with glue.

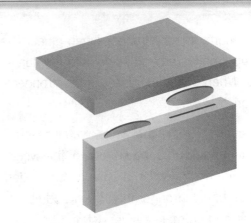

Knock-down Fittings

A wide range of **knock-down fittings** (KD) is now available for DIY and the commercial production markets.

A **cam bolt** is a locking system with a cam that locks the parts together when turned with a screwdriver.

Cabinet screws are used to join kitchen units together.

A **pronged nut** simply taps into a hole to provide a threaded insert that will take a machine screw.

A metal-screwed insert called a **cross dowel** sits in a hole and takes a machine screw.

Modesty blocks are available in a range of colours. These plastic blocks are used to take screws in each direction. They are suitable for simple box joints.

A Cam Bolt

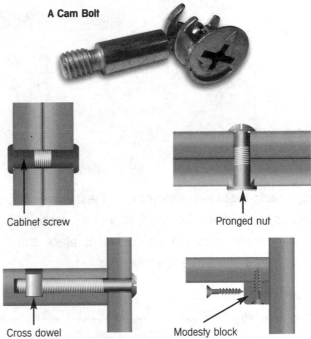

Cabinet screw

Pronged nut

Cross dowel

Modesty block

Quick Test

1. What can be done to make a woodwork joint stronger?
2. Name the wooden pegs used in some joints.
3. What are nails used for?
4. What do the letters KD stand for?

KEY WORDS

Make sure you understand these words before moving on!

- Tenon saw
- Dovetail saw
- Coping saw
- Circular saw
- Bandsaw
- Jigsaw
- Scroll saw
- Butt joint
- Halving joint
- Dowel joint
- Lap joint
- Housing joint
- Mortise and tenon
- Dovetail joint
- Nail
- Screw
- Biscuit joint
- Knock-down fittings
- Cam bolt
- Cross dowel
- Modesty blocks

Adhesives and Finishes

Glue Types

All joints become much stronger when glued. The glue fills up small gaps and soaks into the fibres of the wood. When dried, the glue can be stronger than the original wood.

Polyvinyl acetate (PVA) is a white, water-based adhesive that soaks into the surface of the wood and sets once all the water is absorbed. It's often regarded as being stronger than the wood fibres themselves and so makes a very strong bond. It isn't usually waterproof.

Synthetic resin is a waterproof adhesive which needs to be mixed into a creamy consistency with water. Chemical hardening then takes place and it becomes very hard and brittle.

Hot melt glue is useful for quick modelling, but it's difficult to control so isn't often used in final products.

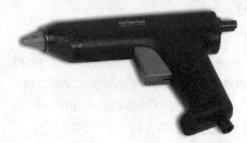

Epoxy resin is a very versatile but expensive adhesive that will stick most clean, dry materials together. Equal amounts of resin and hardener are mixed together and it sets chemically to be very hard.

Contact adhesive is ideal for gluing plastic laminates onto chipboard. Both surfaces are coated and allowed to become touch-dry. Adhesion takes place as soon as the two surfaces meet. The solvent fumes are very dangerous and good ventilation is essential.

Latex adhesive is a rubber solution which is cheap and very safe. This makes it ideal for gluing felt into a box. It does give off a fishy smell, but the fumes aren't dangerous.

Abrasive Papers

Abrasive papers are made by gluing small chips of abrasive material onto a paper sheet. The abrasive material might be **garnet, aluminium oxide** or **glass**.

Each sheet is numbered with the sieve size that the abrasive passed through, so higher numbers are finer abrasives. The papers are also graded into coarse, medium and fine with **flour paper** being the finest. Glass and garnet paper are used mainly for wood.

Abrasive paper is usually wrapped around a cork block to keep it flat. The wood should be glass papered along the line of the grain until only fine lines can be seen. Don't move on to the next finer grade until all pencil marks, etc. have gone.

Finishing Wood

Polish is applied to wood in order to…
- protect it from moisture
- protect it from insect attack
- enhance the colour of the grain
- make it easier to wipe the surface clean.

Wood stains can be used to enhance the colour of the timber and show up the grain patterns. They are usually applied with a cloth.

Sanding sealer / french polish is a solvent-based product similar to a varnish used to seal timber. The quick-drying liquid seals the surface but raises the fibres of the timber so they have to be cut back (rubbed down) with fine abrasive paper. It's suitable as a first coat before applying varnish or wax polish.

Polyurethane varnish is a tough, heatproof and waterproof finish available in different colours and with a matt, satin or gloss surface finish. Three coats are usually applied with a brush then rubbed down with fine glasspaper until the material's surface is smooth.

Cellulose is a hard, quick drying finish that's useful when wood turning.

Wax polish is usually used on wood. Wax fills the porous surface of the timber and a layer of polish is built up on the surface of the material. It can be applied…
- by hand
- with a cloth
- using a buffing wheel.

There are various types of polish suitable for timber including beeswax and silicone polish.

Quick Test

1. What adhesive would you use to glue two pieces of wood together?
2. What adhesive would you use to glue a piece of wood to a piece of metal?
3. Sketch a piece of wood being 'rubbed down' with glass paper.
4. Give two reasons why a surface finish is applied to a piece of wood.

KEY WORDS
Make sure you understand these words before moving on!
- Polyvinyl acetate
- Hot melt glue
- Epoxy resin
- Contact adhesive
- Latex adhesive
- Garnet
- Aluminium oxide
- Glass paper
- Flour paper
- Sanding sealer
- French polish
- polyurethane
- Cellulose
- Wax polish

Practice Questions

1 Give two properties of pine.

a) ....   b) ....

2 Give one use for teak. ....

3 Give two properties of ash.

a) ....   b) ....

4 A dark coloured hardwood is needed for a jewellery project. What would be a suitable wood?
Tick the correct option.

    A   Oak   ⬜   **B**   Mahogany   ⬜

    C   Pine   ⬜   **D**   Blockboard   ⬜

5 Which of the following statements describe valid reasons for applying a wood finish? Tick the correct options.

    A   To help protect the surface from moisture   ⬜

    B   To stop it from being dented   ⬜

    C   To save time when finishing   ⬜

    D   To help protect the wood from insect attack   ⬜

    E   To improve the appearance of the grain   ⬜

    F   To cover up the mistakes that have been made   ⬜

6 The table contains the names of six wood joining methods. Match descriptions **A, B, C, D, E** and **F** with the methods **1–6** in the table. Enter the appropriate number in the boxes provided.

Wood Joints			
1	Wood screws	**4**	Nails
2	Butt joint	**5**	Housing joint
3	Dovetail joint	**6**	Lap joint

    A   A corner joint for boxes, which is strong if glued   ⬜

    B   A joint used for holding up shelves in a cabinet   ⬜

    C   A weak fastening, which only relies on the strength of the glue   ⬜

    D   Permanent fastenings that can be removed with a screwdriver   ⬜

    E   A very strong joint, which can only be pulled apart in one direction and is used for drawer fronts   ⬜

    F   A temporary fastening that can be pulled out again after the glue has dried   ⬜

FREE

Letts
and
LONSDALE

ESSENTIALS

GCSE Design & Technology
Resistant Materials
Controlled Assessment Guide

About this Guide

The new GCSE Design & Technology courses are assessed through…
- written exam papers
- controlled assessment.

This guide provides…
- an overview of how your course is assessed
- an explanation of controlled assessment
- advice on how best to demonstrate your knowledge and skills in the controlled assessment.

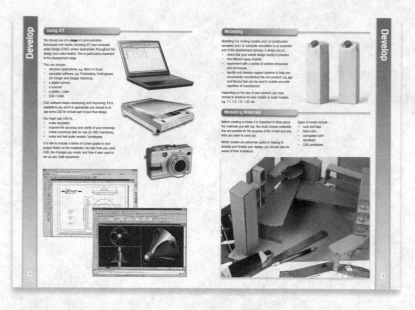

What is Controlled Assessment?

Controlled assessment has replaced coursework. It involves completing a 'design and make' task (two separate tasks for OCR) within a set number of hours.

Your exam board will provide you with a range of tasks to choose from. The purpose of the task(s) is to see how well you can bring all your skills and knowledge together to design and make an original product.

You must produce individual work under controlled conditions, i.e. under the supervision of a teacher.

Your teacher can review your work and give you general feedback. However, all the work must be your own.

How is Controlled Assessment Marked?

Your teacher will mark your work using guidelines from the exam board. A moderator at the exam board will review these marks to ensure that they are fair.

You will not just be marked on the quality of your end product – the other stages of design and development are just as important, if not more so!

At each stage of the task(s), it is essential to clearly communicate…
- what you did
- how you did it
- why you did it

You will be marked on the quality of your communication too.

Contents

This guide looks at the main stages you will need to go through in your controlled assessment task(s), providing helpful tips and advice along the way.

Exam Board	Course	Written Paper	Controlled Assessment
AQA	Full Course	• 2 hours • 120 marks • 40% of total marks Section A (30 marks): A design question based on context, which you will be notified of before the exam. Section B (90 marks): Covers all the content on the specification, i.e. all the material covered in your *Essentials Revision Guide*.	• Approx. 45 hours • 90 marks • 60% of total marks (equivalent to 2 marks per hour)
Edexcel	Full Course	• 1 hour 30 minutes • 80 marks • 40% of total marks	• Approx. 40 hours • 100 marks • 60% of total marks (equivalent to 2½ marks per hour) The 'design and make' activities can be linked (combined design and make) or separate (design one product, make another).
OCR	Short Course and Full Course (Yr 1)	**Sustainable Design:** • 1 hour • 60 marks • 20% of total marks (40% of short course) Section A: 15 short answer questions. Section B: 3 questions requiring answers that may involve sketching, annotation, short sentences or more extended writing.	**Introduction to Designing and Making:** • Approx. 20 hours • 60 marks • 30% of total marks (60% of short course) (equivalent to 3 marks per hour)
	Full Course (Yr 2)	**Technical Aspects of Designing and Making:** • 1 hour 15 minutes • 60 marks • 20% of total marks Section A: 3 questions based on the technical aspects of working with materials, tools and equipment. Section B: 2 questions on the design of products reflecting the wider aspects of sustainability and human use. One of these questions will require a design response.	**Making Quality Products:** • Approx. 20 hours • 60 marks • 30% of total marks (equivalent to 3 marks per hour)

Important Considerations

Unlike your teacher, the moderator will not have the opportunity to see how you progress with the task. They will not be able to talk to you or ask questions – they must make their assessment based only on the evidence you provide. This means that it is essential to communicate your thoughts, ideas and decisions clearly at each stage of the process:

- Organise your folder so the work is in a logical order.
- Ensure that text is legible and that spelling, punctuation and grammar are accurate.
- Use an appropriate form and style of writing.
- Make sure you use technical terms correctly.

Because you only have a limited amount of time, it is essential to plan ahead. The table below gives suggested times for each of the stages.

These times are guidelines only and you should produce your own more detailed time plan. You need to divide the total time for each stage between the individual tasks to ensure that you spend the majority of your time working on the areas that are worth the most marks.

That doesn't mean that the other tasks aren't important, but quality, rather than quantity, is key.

You should aim to produce about 20 x A3 sheets, or equivalent, for your folder (10 for a short course or for separate design and make tasks).

*AQA award up to 6 marks for clarity of communication throughout your folder. Whilst these marks are important, 84 of the total 90 marks are for the content, so make good use of your time – don't waste time creating elaborate borders and titles!

**Refer to the course specification for a more detailed breakdown of how marks are allocated for the OCR design and make tasks.

Stage	Tasks	AQA Marks	AQA Guideline Time (Hr)	Edexcel Marks	Edexcel Guideline Time (Hr)	OCR A561** Marks	OCR A561** Guideline Time (Hr)	OCR A563** Marks	OCR A563** Guideline Time (Hr)
Investigate	Analysing the brief	8	4	15	5	24	8	16	5½
				20	7				
	Research								
	Design specification								
Design	Initial ideas	32	16	15	8				
	Reviewing ideas								
Develop	Developing ideas								
Plan	Product specification			6	2	28	9½	36	12
	Production plan			38	16				
Make	Making product	32	16						
Test and evaluate	Testing and evaluation	12	6	6	2	8	2½	8	2½
Communicate*	Clarity of communication	6	3						
Total		90	45	100	40	60	20	60	20

Analysing the Task

To get the maximum marks, you need to…
- analyse the task / brief in detail
- clearly identify all the design needs.

It is a good idea to start by writing out the task / brief…
- as it is written by the exam board
- in your own words (to make sure you understand what you're being asked to do).

You then need to identify any specific issues that you need to consider before you can start designing the product.

You do not need to write an essay. You could use…
- an attribute analysis table
- a mind map
- a spider diagram
- a list of bullet points.

At this stage it is a good idea to…
- eliminate all the things that you don't need
- make a list of 'Things I need to know'.

Ask yourself the following questions:

- Who will use the end product?
- What will it be used for?
- Where will it be used?
- What sort of shop / retail outlet will it be sold through?
- Are there any cost restrictions that will influence my design?
- How many products would be made if it went into commercial production?

N.B. You need to have clear answers to all of the above before you start designing your product.

Research

Because you don't have very long to conduct your research, you need to make sure it is all relevant.

It should help you to make decisions about all the issues that you identified in your product analysis, so these are the areas to focus on.

Make sure you keep accurate records. You will need to refer back to the information throughout the task.

You should know about the different research methods used in commercial design, but be aware that they may not be appropriate for your design task because of the limited time available to you.

Possibly the most useful type of research that you can carry out in the limited time available to you is to interview the client / end-user to find out what they want from the product.

Questions for client / end-user:

- What product do you want?
- What do you want the product to do?
- How and where will you use it?
- What size do you want?
- What style do you want?
- What do you like and dislike about existing products?

Product Analysis

In the limited time available to you for the controlled assessment you are unlikely to be able to carry out detailed analysis and disassembly of existing products.

However, it is a good idea to look at some existing products with your client / end-user to find out…
- what they like about them
- what they don't like about them
- whether or not they are good value for money
- what improvements they would like to see.

You should also consider…
- the life cycle of the product
- whether the product can be recycled

- the effect of the product on our lifestyle
- whether the product is inclusive (or whether some groups of people will not be able to use it).

The purpose of this is to help you produce a product that is better than those already available. It should help you to identify…
- desirable / successful features (features you could incorporate into your design)
- undesirable / unsuccessful features (features to avoid using in your design)
- areas for improvement (areas that you should try to improve upon in your design, e.g. reducing cost, making the product sustainable).

Interviews, Questionnaires and Surveys

Interviews, questionnaires and surveys normally rely on a large sample group to produce reliable data.

You will need to adapt these methods for your design task, e.g. a single client interview or a simple

questionnaire at the end of the task to evaluate your outcome.

This is fine, as long as you show that you understand the pros and cons of doing this in your evaluation.

Research Summary

It is essential to summarise your conclusions and clearly explain how the data gathered through your research will assist you. You should record…
- what you did
- why you did it

- what you hoped to find out (i.e. what your expectations were)
- what you actually found out
- how these findings will affect your design ideas.

Client Interview and Questionnaire

Conclusion

Initial Ideas

Generating ideas is an important part of any design process, and you should allow yourself plenty of time for this stage. This is your chance to show off your creative skills, but make sure your ideas…

- are **realistic** and **workable**
- address **all** the essential criteria on your design specification.

Don't panic if your mind goes blank – try highlighting key words in the original brief to help focus your thoughts. Then use word association to create…

- a mind map
- a spider diagram
- a list of ideas.

You can then play around with some of the words and ideas you came up with. Here are some methods you could experiment with:

- **Drawing through a window** – select an object of interest (nature is a good source) and focus on one part of it to produce interesting shapes and patterns.

- **In the style of**… – borrow elements from past design movements, e.g. Bauhaus, Art Deco, Shaker.
- **Setting rules** – set yourself rules, e.g. you can only use straight lines and two circles in your design.
- **Look at famous designers** – look at the work of famous designers, past and present, for inspiration.
- **Try modelling** – use materials like art straws and card, plasticine, wire and scraps of fabric to experiment with 3D forms.
- **Working with grids** – use squared paper or cut-out symmetrical shapes and experiment with repeated patterns.
- **Reverse engineering** – try making something that already exists by working backwards. Once you fully understand how it is made then you can start to suggest changes.

Presenting Ideas

You need to present your initial ideas clearly, but remember there are no marks for making it look 'pretty'.

To communicate your design ideas clearly and show how they relate to the criteria on your design specification, try using sketches with notes and annotations.

Don't worry about how good your drawings or models are at this stage; it is the variety and feasibility of the ideas that are important.

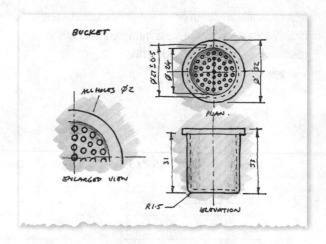

Reviewing Ideas

You need to review your initial ideas in order to select one or two to develop further.

They must satisfy the essential criteria on your design specification, but you will also want to consider…

- which designs satisfy the most desirable criteria
- which designs are most unique / innovative
- which designs are most appealing / attractive.

Ask your client / end-user for their opinion – which ones would they buy?

Developing Ideas

Development is another important part of the design process. Your aim at this stage is to modify and revise your initial idea(s) until you reach the best possible design solution.

When your teacher and the moderator look at your development sheets they will expect to see a design that is **significantly different** and **improved** compared to your initial idea.

To get you started, one method you could try is to draw / model your initial idea, then draw / model it again making one change. Continue drawing or modelling your idea making one change each time and it will evolve into a whole sheet of drawings / photos of models.

To help you make the necessary modifications…
- use tests to ensure that your final design meets all the essential criteria on your design specification
- ask your client / end-user for feedback.

Once you feel that you have reached the best possible design solution, make sure it is presented in a way that someone else can understand.

At this stage, you need to use your knowledge of a wide range of materials, components and manufacturing processes. Your work should show a good understanding of…
- properties of materials and / or components
- the advantages / disadvantages of materials and / or components
- the advantages / disadvantages of manufacturing processes.

This means you must select the most appropriate materials, components and manufacturing processes for your product and justify your choices.

At the end of this process you should have enough information to produce a detailed product specification and / or manufacturing specification.

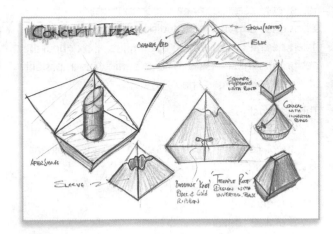

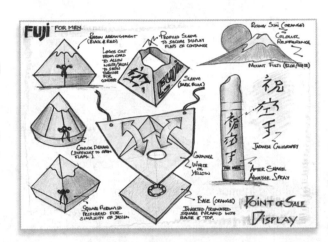

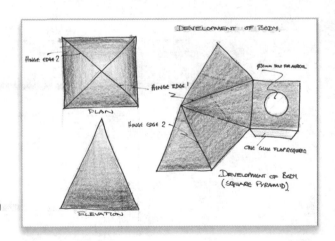

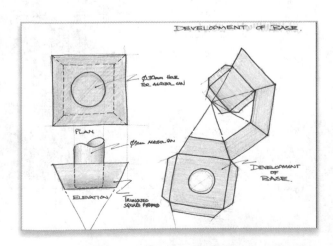

Using ICT

You should use of a **range** of communication techniques and media, including ICT and computer aided design (CAD), where appropriate, throughout the design and make task(s). This is particularly important at the development stage.

This can include…
- standard applications, e.g. Word or Excel
- specialist software, e.g. ProDesktop, ProEngineer, 2D Design and Google SketchUp.
- a digital camera
- a scanner
- a plotter / cutter
- CAD / CAM.

CAD software keeps developing and improving. If it is available to you and it is appropriate, you should try to use some CAD for at least part of your final design.

You might use CAD to…
- make templates
- improve the accuracy and clarity of your drawings
- create numerical data for use on CNC machinery
- make and test scale models / prototypes.

It is vital to include a series of screen grabs in your project folder, so the moderator can see how you used CAD, the changes you made, and how it was used to set up any CAM equipment.

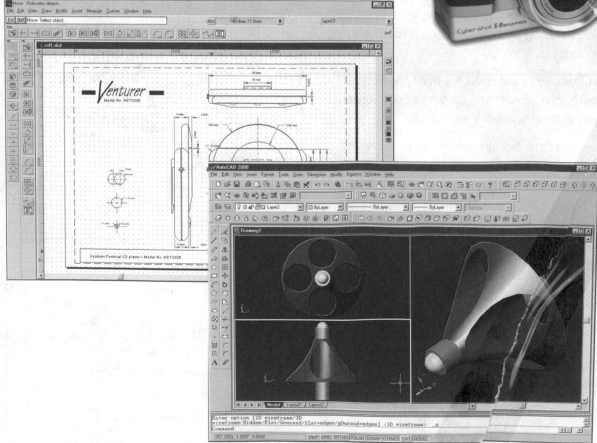

Modelling

Modelling (i.e. making models and / or construction samples) and / or computer simulation is an essential part of the development process. It allows you to…

- check that your overall design works in practice
- trial different types of joints
- experiment with a variety of suitable processes and techniques
- identify and develop support systems to help you successfully manufacture the end product, e.g. jigs and fixtures that can be used to enable accurate repetition of manufacture.

Depending on the size of your product, you may choose to produce full size models or scale models, e.g. 1:1, 1:2, 1:5, 1:10, etc.

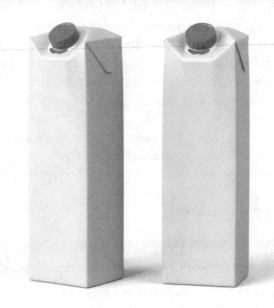

Modelling Materials

Before creating a model, it is important to think about the materials you will use. You must choose materials that are suitable for the purpose of the model and any tests you want to carry out.

Whilst models are extremely useful in helping to develop and finalise your design, you should also be aware of their limitations.

Types of model include…

- card and tape
- foam core
- corrugated card
- styrofoam
- CAD prototypes.

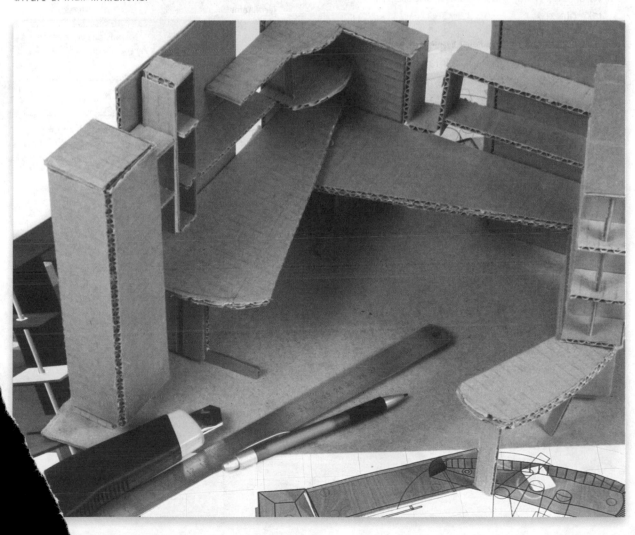

Product / Technical Specification

Your product specification will be more detailed than your design specification. It should include…

- working drawings (drawn, sketched or CAD) – orthographic or isometric drawings, showing dimensions and main construction details
- a cutting list
- details of jigs, fixtures and processes to ensure accurate working especially in multiple production, e.g. to ensure that the four corners of a box, chair or unit are all identical
- a risk assessment.

You should use your product specification to put together a production plan and build and test a prototype of your end product.

Risk Assessment

Look at each process in turn and make a list of possible health and safety risks.

Work back through the list and plan how you will minimise the risks, for example…

- by wearing safety equipment
- by ensuring you know how to use the tools correctly.

Production Plan

Your production plan should show…

- the different stages of manufacture in the correct order
- when and what quality control checks will take place.

A flow chart might be the best way of presenting your production plan, although sometimes a simple chart listing the stages and the equipment that will be used is equally suitable.

If you draw a flow chart, there are different, specific symbols for each stage of the process. The symbols are linked together by arrows to show the correct sequence of events.

You should aim to keep your flow chart as clear and simple as possible.

Flow Chart Symbols

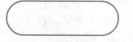

Terminator
Represents start, restart and stop.

Process
Represents a particular instruction or action.

Decision
Represents a choice that can lead to another pathway.

Input / Output
Represents additions to / removals from a particular process.

Flow Chart

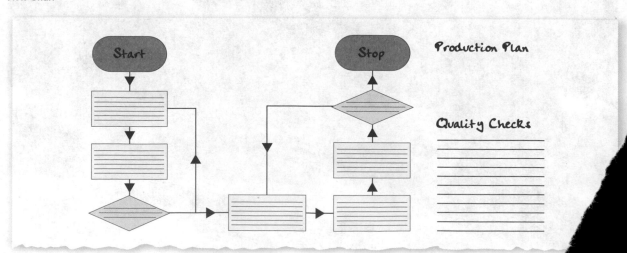

Production Plan

Quality Checks

Manufacture

Your revision guide includes information on many of the materials, tools, processes and methods relevant to your particular subject.

In making the prototype of your final product you should demonstrate that, for each specific task, you can correctly select and use all of the following safely:

- Appropriate tools and equipment
- Appropriate processes, methods and techniques (including CAD / CAM where relevant)
- Appropriate materials

You can do this by carrying out the necessary practical processes safely and with **precision** and **accuracy**.

Remember, all the materials, methods and processes that you choose must help to make your product the best possible design solution for the brief. Don't include something just to show off your skills!

The finished product should be...

- accurately assembled
- well finished
- fully functional.

Don't worry if it doesn't turn out quite the way you hoped though – you will earn marks for all the skills and processes you demonstrate, so make sure you record them all clearly in your folder.

*For each stage of production, you might want to include...

- a list and / or photograph of materials used
- a list and / or photograph of tools used
- a flow chart / step-by-step description of the process carried out
- an explanation of any safety measures you had to take.

N.B. There are no marks available from AQA for doing this.

Industry

You should have a good understanding of the methods and processes used in the design and manufacturing industries in your subject area.

Although you will probably only produce one final product, it is important to show that you are aware of various possible methods of production and how your product would be manufactured commercially. You should explain this in your project folder.

If your product could potentially be manufactured using several different methods, try to list the pros and cons for each method and then use these lists to make a decision about which method you would recommend.

If you know that a method or process you are using to make your product would be carried out differently in a factory, make a note of this in your project folder – this will show your teacher and moderator how much you know!

Quality Control and Assurance

Your revision guide looks at some of the quality control tests and quality assurance checks used in industry that are relevant to your particular subject (see p.23–24).

Using a jig is one way of introducing quality control in the manufacture of your own product.

Any quality checks you need to make should be included in your flow chart / production plan.

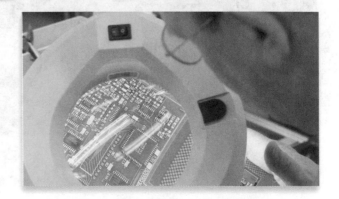

Testing

A **range** of tests should be carried out to check the performance and / or quality of the final product. You need to justify each test you carry out, i.e. explain why it is important.

Tests do not have to be complicated. They just need to be sensible and helpful, e.g. test the usability and functionality of the end product.

When working with food, taste tests are often the best option.

Keeping records is very important. In your project folder you need to…
- explain what tests were carried out
- explain why the tests were carried out
- describe what you found out
- explain what modifications you would make, based on the test results
- include a photo of the prototype in use.

You should not test your prototype to destruction, but it is a good idea to take photos of your product before testing begins just in case anything goes wrong.

Companies undertake numerous tests on prototypes before a product goes into mass production. These can sometimes include testing to destruction. If you think this type of testing is necessary, you should do it at the development stage using models – the moderator will want to assess your prototype so it needs to be intact!

5kg weight

10kg weight

15kg weight

Evaluation

Evaluation is an ongoing process. During the design and development process, every decision you made (providing it is clearly justified) and all the client / end-user feedback will count towards your evaluation.

The final evaluation should summarise all your earlier conclusions and provide an objective evaluation of the final prototype.

When carrying out an evaluation, you should…
- refer back to the brief
- cross-check the end product against the original specification
- obtain client comments and feedback
- take a photograph of the client using the product
- carry out a simple end-user survey.

You need to establish…
- whether the product meets all the criteria on the original brief and specification
- whether the product is easy to use
- whether the product functions the way it was intended to
- what consumers think of the style of the product
- whether consumers like / dislike any features
- whether consumers would purchase the product and what they would be prepared to pay for it
- what consumers think the advantages and disadvantages are compared to similar products
- what impact the making and using of the product has on the environment.

Depending on what you find out, you can include suggestions for further modifications in your evaluation.

Honesty is the best policy when writing evaluations. If something didn't work, say so – but always suggest a way of preventing the same problem from occurring in the future.

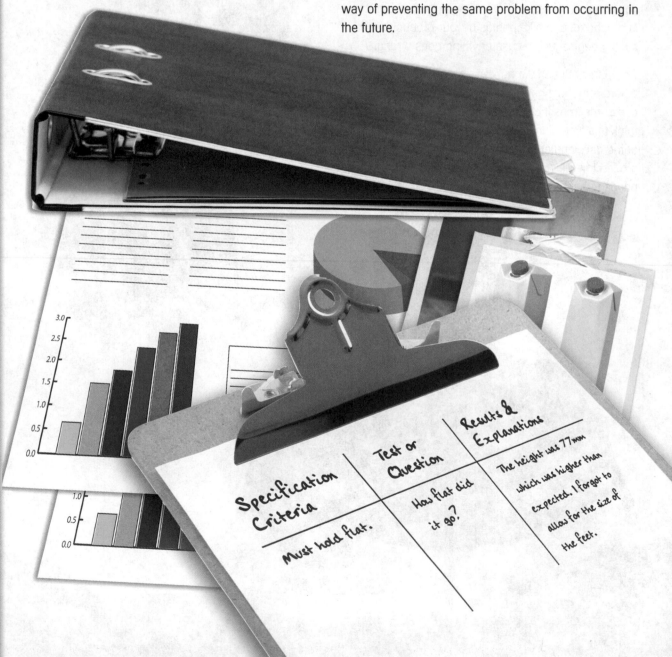

7 (Circle) the correct options in the following sentences.

 a) A **firmer** / **bevel-edged** chisel is used when cutting an angle of less than 90°.

 b) A technique called **vertical** / **horizontal** paring is used to clean out a housing joint that is held in a vice.

8 Fill in the table to show which adhesives would be suitable for the following situations.

Adhesive	Situation
a)	Gluing a piece of indoor furniture together.
b)	Gluing an outdoor table for use in the garden.
c)	Gluing the lining into a jewellery box.
d)	Gluing a plastic laminate onto a piece of MDF.

9 What do each of the acronyms below stand for?

 a) MDF _____   **b)** PVA _____

10 Tick the options below which show the most suitable material that would be used.

 a) A material used to make cheap kitchen cabinets.

  A Pine ☐   **B** MDF ☐

  C Oak ☐   **D** Ash ☐

 b) A material used to make garden furniture.

  A Hardboard ☐   **B** Parana pine ☐

  C Teak ☐   **D** Mahogany ☐

11 Choose the correct words from the options given to complete the following sentences.

plywood   **hardwood**   **origins**   **pine wood**   **tropical rainforest**   **sustainable**

random   **two for one**   **on their own**

When choosing a suitable _____ to make a jewellery box for a client, the designer

needs to consider the _____ of the wood. It shouldn't come from a

_____, but should have been grown in a _____ way, where new

saplings are planted (_____) and the felling is controlled.

Metals and their Properties

Metals and their Properties

Metal ore is mined from the ground, and the metal is then extracted from the rocks. There are three main types of metal: **ferrous**, **non-ferrous** and **alloys**.

Ferrous metals consist of iron, carbon and other elements. Example are wrought iron, mild steel, tool steel, stainless steel and cast iron. Most ferrous materials are prone to **rusting** and can be picked up with a **magnet** (the exception is stainless steel, which is designed not to rust and some grades are non-magnetic).

Non-ferrous metals don't contain any iron, so they aren't attracted to a magnet and don't rust when exposed to moisture (but they do tarnish and oxidise). Examples are copper, aluminium, tin and zinc.

Alloys are examples of **combined materials**. Alloys are metals that contain two or more elements. The elements that make up an alloy may be metal or non-metal. Alloys are chosen for a particular purpose because of their properties, for example their low melting points, corrosion resistance or overall weight. Examples are casting alloy, pewter and brass.

There are factors to consider when choosing a metal for a specific purpose:

- **Workability** – some metals are much easier to work with than others.
- **Structural strength** – metals vary from weak to very strong.
- **Appearance** – metals differ greatly in colour and appearance.

For example, you should consider: elasticity, ductility, malleability, hardness and work hardness, brittleness, toughness, tensile strength, compressive strength and resistance to rusting / tarnishing.

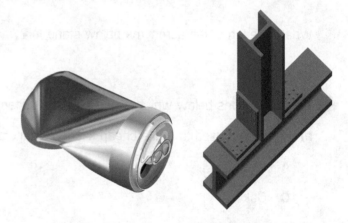

Heat Treatment

The crystalline structure of a metal affects how it reacts to working and cutting. The size of the grains is important and can be changed by **annealing** (heating and cooling the metal in a controlled way) or softening the metal. By hardening and annealing metals we can make them...

- malleable enough to form into shapes
- hard enough to cut all other metals.

If a metal becomes deformed by working (hammering or bending) it becomes **work hardened** and continuing to work the metal will make it brittle. If the metal is annealed...

- the heating allows the grains to re-form
- on cooling, the metal will have returned to its original condition and working can continue.

When a **ferrous** metal is annealed it becomes very soft and workable. It can be annealed in the following way:

1. Heat it to bright cherry red (725°C) and soak it in the heat for a few minutes.
2. Cool it very slowly (to allow the grains to grow).

A **non-ferrous** metal can be annealed in the following way:

1. Heat it to dull red (500°C).
2. Either quench in cold water or allow it to cool slowly.

The scale or oxide on the surface of the metal can be cleaned off with acid or an abrasive.

Metals and their Properties

Hardening

If high carbon or tool steel is **hardened**, it is at its maximum hardness and able to cut other steels. Steel can be hardened in the following way:

1. Heat the metal to cherry red (720°C).
2. Soak the metal in the heat for a few minutes.
3. Quickly quench it in water or oil.

Tempering

Hardening makes steel very hard and means that it will not wear away, but it leaves it brittle and liable to crack.

Tempering reduces the hardness a little until a workable balance is found between hardness and brittleness (toughness).

Steel can be tempered in the following way:

1. Clean the metal so its colour can be seen.
2. Reheat the steel until it's between 230° and 300°C depending on the toughness you need for the metal (see colour chart).
3. As soon as the correct colour is seen, stop the heating and quench the metal.

Mild steel doesn't have enough carbon in it to be hardened, but it can be **case hardened**, i.e. given a thin outer layer of hardened steel by soaking it in a bed of red hot carbon for several hours at 950°C.

Colour	Hardness	Temp.(°C)	Uses
Pale straw	Hardest	230	Lathe tools and scribers
Straw		240	Drills and milling cutters
Dark straw		250	Taps and die punches
Brown	Tough and hard	260	Plane irons, shears and chisels
Brown / purple		270	Scissors and knives
Purple		280	Cold chisels and saws
Dark purple		290	Screwdrivers
Blue	Toughest	300	Springs, spanners and needles

Quick Test

1. Name the three main metal groups.
2. Give three properties that might influence a designer in their choice of a metal.
3. What is the colour of a tempered steel spring?
4. Why is copper annealed during the forming process?

51

Ferrous Metals

Name	Description	Uses
Cast iron	• Re-melted pig iron with some small quantities of other metals • Typically 93% iron with 4% carbon • Very strong in compression, but brittle	• Metalwork vices • Brake discs and drums • Car cylinder blocks • Manhole and drain covers • Machinery bases
Mild steel	• Iron mixed with 0.15–0.3% carbon • Ductile and malleable • Rusts very quickly if exposed to moisture	• Nuts • Bolts • Car bodies • Furniture frames • Gates • Girders
Tool steel	• Also known as 'medium' or 'high carbon' steel • Up to 1.5% carbon content • Strong and very hard	• Hand tools, e.g. chisels, screwdrivers, hammers, saws • Garden tools • Springs
High speed steel	• Contains a high content of tungsten, chromium and vanadium • Brittle but resistant to wear • Used in machining operations where high speeds and high temperatures are created	• Drill bits • Lathe tools • Milling cutters
Stainless steel	• An alloy of iron with typically 18% chromium and 8% nickel • Very resistant to wear and corrosion • Doesn't rust	• Kitchen sinks and general fittings in commercial kitchens • Cutlery • Dishes • Teapots • Surgical instruments

Non-ferrous Metals

Types of Non-ferrous Metals

Name	Description	Uses
Aluminium	• Light grey • Can be polished to a mirror-like appearance • Light in weight • Can be anodised to protect the surface and give it colour	• Cooking foil • Saucepans • Chocolate wrappers • Window frames • Toy cars • Ladders
Copper	• Reddish-brown, but can turn green after exposure to oxygen • Ductile and malleable • An excellent conductor of heat and electricity	• Plumbing and electrical components • Domed roofs (copper covered)
Lead	• A heavy metal with a blue-grey surface • Soft and malleable • Has a high resistance to corrosion from moisture and acids	• Car battery cells • Weather proofing for buildings and around chimneys • Plumber's solder
Tin	• Bright silver • Ductile and malleable • Resistant to corrosion • Tinplate is steel with a tin coating	• Most commonly used as a coating on food cans and similar packaging
Zinc	• Very weak • Extremely resistant to corrosion from moisture	• Used as a coating on steel buckets, screws and roofing sheets (galvanised steel) • Die casting alloys
Gold and silver	• Precious metals • Very ductile and malleable • Silver tarnishes, but gold isn't affected by oxidation	• Jewellery • Plated onto electrical wires to improve contact and reduce resistance

Alloys and Marking Out Metals

Types of Alloys

Name	Description	Uses
Brass	• Hard, yellow metal • An alloy of about 65% copper and 35% zinc • Often cast and machined, then chromium plated	• Decorative metal work, e.g. door handles, candlesticks and boat fittings • Plumbing accessories
Guilding metal	• An alloy of about 85% copper and 15% zinc • Has a much darker colour than brass	• Architectural metalwork and jewellery • Often used in sheet-metal sculptures
Pewter	• Now a lead-free alloy (for safety) that can be easily cast • Made of 92% tin, 6% antimony and 2% copper • Polishes to a bright, mirror-like finish • Low melting point	• Drinking tankards • Jewellery • Picture frames • Decorative gifts
Casting alloy (LM4)	• Mainly aluminium with 3% copper and 5% silicon • Looks like pure aluminium	• Sand casting • Die casting engine components, especially on motorcycles
Duralumin	• An aluminium alloy • Almost as strong as steel but only 30% of the weight • Mainly aluminium with 4% copper and 1% manganese and magnesium	• Aircraft bodies • Cars • Door handles

Measuring Lengths and Angles

A **steel rule** is used to measure lengths and to set odd-legs and dividers.

Engineering Square

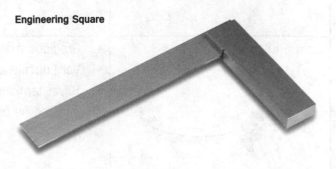

An **engineering square** is used to...
• mark lines at exactly 90° from the edge of the material
• to check square components or corners.

An **engineering bevel** is an adjustable angle marker that can be set to any angle.

Marking and Joining Metals

Marking the Surface

There are several tools for marking surfaces:

- A **scriber** is a sharp tool used for marking accurate lines.
- Marking **blue** is thin ink spread on a metal's surface and left to dry. When scratched it leaves a visible, shiny line.
- **Dividers** are used to mark out circles or arcs. One point digs into the material whilst the other point scribes a line into the surface of the material.
- **Odd-leg calipers** are used to scribe lines parallel to a straight edge. The notched leg slides along the edge of the metal while the scriber point marks a line.
- A surface gauge (or **scribing block**) and **surface plate** are used for marking parallel lines accurately.
- A **dot punch** has a conical point ground to 60° that is used for marking the positions of drill holes. It's also used to make small dotted indentations on lines that have been marked out.

This makes the lines clearer when cutting or shaping the material by hand or with a machine tool.

- A **centre punch** has a conical tip ground to 90° that is used for marking the centre of holes for drilling. The punch is put into the dot punch mark and enlarges it. The punch leaves a small indentation which prevents the drill bit from wandering across the surface.

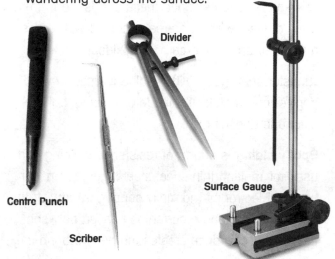

Divider

Surface Gauge

Centre Punch

Scriber

Joining Metals

Heat can be used to join metals together:

- **Welding** uses heat to melt the edges of the metals being joined. This forms a pool of molten metal that when cooled forms the joint (like casting).

- **Soldering** uses a bonding alloy to form a joint (like gluing). The 'parent metals' don't melt. The surface is cleaned with flux and heat is then applied using a gas torch. The solder is applied, which melts and 'runs' between the two parent metals.

Quick Test

1. Give three reasons why a designer would choose to use gold for jewellery.
2. Why is tin used in the food industry?
3. What is engineer's blue used for?
4. Why does an engineer use two different punches for marking out a hole?
5. What is the difference between welding and soldering?

KEY WORDS

Make sure you understand these words before moving on!

- Engineering square
- Engineering bevel
- Scriber
- Blue
- Dividers
- Odd-leg calipers
- Surface plate and Scribing block
- Dot punch
- Centre punch
- Welding
- Soldering

Joining Metals

Welding

Gas welding uses an acetylene torch to heat up the joint. A mixture of acetylene gas and oxygen produces a very small, hot flame which melts both the filler rod and the surrounding metal.

Metal inert gas (**MIG**) welding is often used in schools to weld steels. An electrical spark creates the heat by arcing between the electrode and the work-piece. The electrode melts and is fed in from a roll of steel wire. The area is cooled with a gas mixture of argon and carbon dioxide.

Tungsten inert gas (**TIG**) welding is used to weld non-ferrous metals. It's like MIG welding, but the electrode doesn't melt.

Spot welding is a form of resistance welding and is used for melting thin sheet steel together, e.g. car bodies. Electrodes (normally copper) sandwich the metal together and a current is passed between them. The resistance creates the heat to bond the two metals in a tiny spot.

Gas Welding

Spot Welder

Visor and Gloves

Health and safety: You must always wear a visor or protective facemask when welding to protect your eyes.

Soldering

Soft soldering joins metal parts together using a lead-free alloy. It's used for light applications. e.g. electrical connections and plumbing joints:
1. Flux is applied to the join to clean it.
2. The fluxed joint is heated using a gas torch or metal soldering bit.

Hard soldering or **brazing** is used for heavier applications as the joint is much stronger. The brass bonding alloy (or **spelter**) melts at a much higher temperature than soft solder. It's used for joining mild steel, but copper can also be brazed.
1. A **borax flux** is mixed to a paste with water and applied to the join to clean it.
2. It's heated to an orange colour with a gas torch and the spelter melts around the join.

Silver soldering is like brazing, but it uses a silver-based alloy. It's used on brass, copper and guilding metal as the silver alloy melts at a lower temperature than brazing spelter.

Nuts and Bolts

Mechanical joints…
- allow different materials to be joined to metals
- can be dismantled for repair or maintenance.

There are many different varieties of **nuts**, **bolts** and **washers**. Threads are sometimes cut into one of the metal pieces instead of using a nut.

Bolts are made with many different heads including…
- **countersunk** head – tightened with a screwdriver
- **cheese head** – tightened with a screwdriver
- **hexagonal head** – tightened with a spanner.
- **socket head** – tightened with an allen key.

Threads also vary in size, but metric threads are now almost standard in schools (M3 to M12 are the most popular). They are available in many lengths, typically 20–100mm. Smaller bolts are called machine screws and have the thread over the entire length.

Nuts need to match the same thread as the bolt. **Wing nuts** are tightened by hand and are useful for temporary joints. Hexagonal heads are tightened with a spanner.

A washer is usually used under the nut to spread the pressure and protect the surface of the materials being joined together. This might be a plain ring washer. A spring washer is often used to stop the nut vibrating loose.

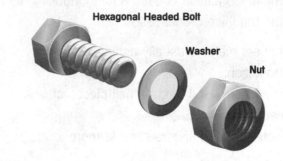

Hexagonal Headed Bolt

Washer

Nut

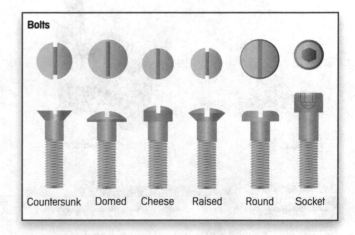

Bolts

Countersunk Domed Cheese Raised Round Socket

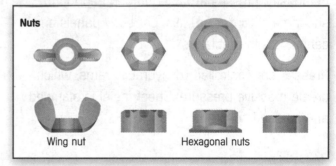

Nuts

Wing nut   Hexagonal nuts

Rivets

Rivets are used to make a more permanent joint than nuts and bolts. They hold the material together by forming a head on both sides of the material. Traditionally, the blank end of the rivet is hammered to form a second head, but **pop rivets** are now more common. They are fitted using a rivet gun from one side of the material.

Rivets come with a variety of heads. They are normally made from soft mild steel, copper or aluminium.

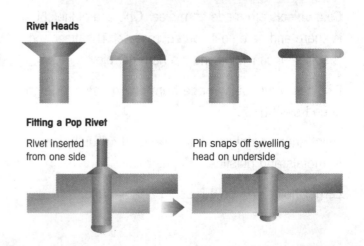

Rivet Heads

Fitting a Pop Rivet

Rivet inserted from one side

Pin snaps off swelling head on underside

Cutting Metals

Shearing and Bending

Sheet metal fabrication is used a lot in industry for hand and machine processes.

Cutting sheet metal is usually achieved using a **shearing** action:

- **Tinsnips** are used to cut small pieces of sheet metal.
- **Bench-mounted** shears provide more leverage.

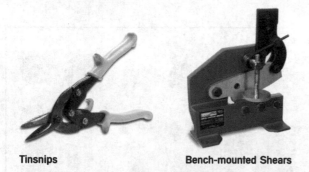

Tinsnips　　　　　**Bench-mounted Shears**

Bending sheet metal can be done in several ways. Folding bars are a common method, but many schools have bending machines.

Folding Bars In a Vice　　　　**Sheet Metal Folding Machine**

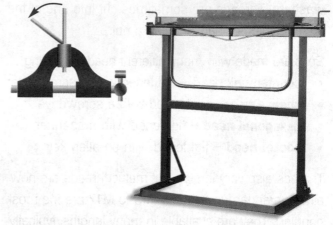

Pressing

In industry, presses are used to bend and form sheet metal (**pressing**), e.g. car body panels and central heating radiators.

Presses are controlled by hydraulic rams, which create massive pressure. Sheet metal is stamped and pressed cold.

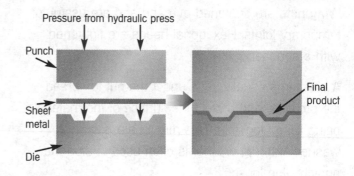

Pressure from hydraulic press

Punch

Sheet metal

Die

Final product

Drilling and Chiselling Metals

Cold chisels are made from steel. One end of the chisel is sharp and hard so it can cut material. The other end is soft, so it can withstand the hammer blows.

Drill bits are usually made from carbon steel or high speed steel (HSS).

Twist drills or jobbers drills are used for drilling holes in metals and plastics.

Larger holes can be cut using a tank cutter.

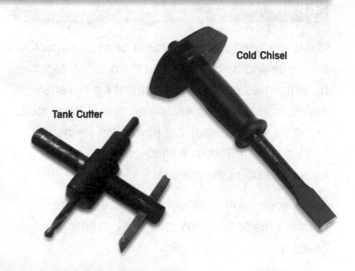

Cold Chisel

Tank Cutter

Screws

Screw cutting can be carried out on most metals. A tapping hole (smaller than the thread) must be drilled in the metal before an internal thread can be cut.

Drill size

Internal, or **female**, threads are usually cut (**tapping**) with a **tap**, held in a tap wrench. A tap is a very hard steel tool which makes its own thread as it's twisted into a hole drilled into the material.

Threading is the cutting of an external, or **male**, thread. The tool used is called a split **die** and is held in a die stock so that it can be turned.

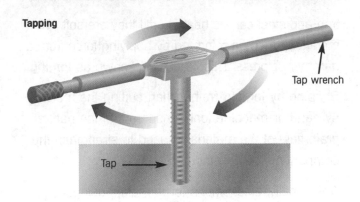

Tapping

Tap wrench

Tap

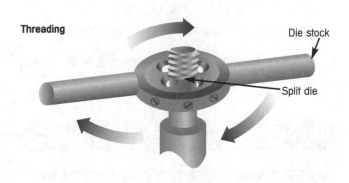

Threading

Die stock

Split die

Thread Cutting

Threads can be cut with a **lathe**.

In industry, many threads are rolled. Hard-threaded rollers rotate the material and press it into shape. This is a form of **cold forging**.

Quick Test

1. What do the letters MIG stand for?
2. Name the system used to weld car bodies.
3. What would you use to make a 5mm hole in a piece of aluminium?
4. What is a tap used for?
5. What is a split die held in?

KEY WORDS

Make sure you understand these words before moving on!

- Gas welding
- MIG
- TIG
- Soft soldering
- Brazing
- Spelter
- Borax flux
- Silver soldering
- Nuts
- Bolts
- Washer
- Wing nuts
- Rivets
- Pop rivets
- Shearing
- Bending
- Female
- Tapping
- Threading
- Male

Cutting Metals

Hand Forging

Iron and steel can be heated until they are soft. The metal can then be reformed by applying force from a hammer or press. This process is known as **forging**.

Shaping by forging (rather than cutting the material by hand or machine tool), ensures that the natural grain flow of the material is used to strengthen the components being made.

Traditionally, blacksmiths forge metals using…

- a hearth to heat the iron or steel
- an anvil to withstand the hammer blows.

By bending, twisting and hammering, a wide variety of forms can be produced.

Drop Forging

Industrial forging is done by a process called **drop forging** or **die forging**:

1. A piece of white hot metal is placed between two dies.
2. A very large force is applied in a single blow by a mechanical hammer.

A similar cold forging process is used to stamp out coins and medals. It is appropriately called **coining**.

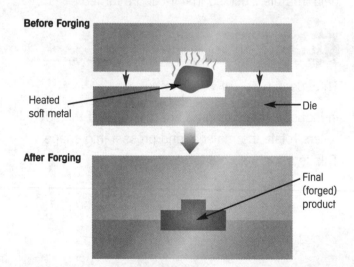

Before Forging

Heated soft metal

Die

After Forging

Final (forged) product

Sawing

Sawing is one of the oldest methods of cutting materials. Teeth are triangular and shaped so that they remove a small amount of material on the forward stroke. The blade is waved to make a cut wider than the blade to reduce friction.

As a general guide, at least three teeth should be on the material at any time. Metal cutting saws have small teeth on a blade supported in a frame.

Hacksaw Blade

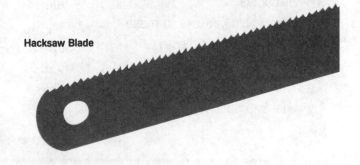

Hand Saws

The **hacksaw** is the general workshop saw for all metals:

- Teeth are available in different **teeth per inch (TPI)** with 20TPI being usual for bench work.
- The blade is tensioned by turning a wing nut and can be changed when worn down.

The **junior hacksaw** is the small workshop saw for thin metals:

- Teeth are usually 32TPI (finer than the hacksaw).
- The blade is tensioned by the spring in the steel frame and can be changed when worn down.

The **piercing saw** is a very fine toothed saw:

- It's used for detailed or accurate work, e.g. jewellery making.
- Blades are held in place by two small wing nuts.

Hacksaw

Junior Hacksaw

Machine Saws

The **powered** hacksaw...

- uses a forwards / backwards motion that copies the manual version
- is driven by a crank slider mechanism.

Scroll saws use blades that can be used for thin sheet metals.

Quick Test

1. Why are forged metals stronger than cut metals?
2. What do the letters TPI stand for?
3. Why is a hacksaw blade slightly 'wavy' in plan view?
4. How many teeth should be in contact with the material at any time?
5. What saw would you use for fine jewellery cutting?

KEY WORDS

Make sure you understand these words before moving on!

- Forging
- Drop forging
- Coining
- Hacksaw
- TPI (teeth per inch)
- Junior hacksaw
- Piercing saw

Casting

Split Pattern Casting

Casting involves pouring metal, which has been heated and melted, into a mould. Any waste material can be re-melted and used again (recycled). **Sand casting** is used to shape metals such as cast iron, aluminium and brass.

The following method, which is common in schools, is almost identical to the industrial process.

1. A **pattern** is made from a timber, e.g. MDF or Jelutong. The pattern is made in two halves and attached to a board.
2. The pattern is sandwiched between open boxes called a **cope and drag**.
3. A special, oil-bound sand is used to fill each box. One of the boxes also contains two tapered wooden pegs or **sprues**, which form the **pouring spout** (**runner**) and the **riser** to let the air out.
4. The pattern is removed and the space left is filled with molten metal.

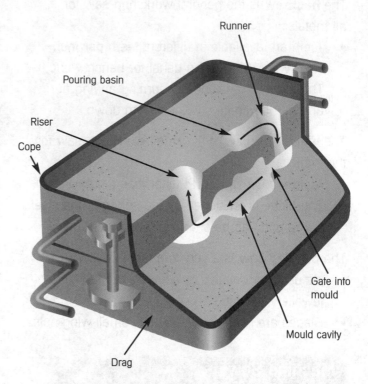

Runner
Pouring basin
Riser
Cope
Drag
Gate into mould
Mould cavity

Lost Pattern Casting

Lost pattern casting is used to form metals, e.g. aluminium, into complete forms that would not be possible with split pattern casting.

The following method is used:

1. A pattern is made from polystyrene foam.
2. The foam is buried into the sand and a runner and riser added.
3. The molten metal is poured into the mould and instantly burns away the polystyrene foam. The space left is filled with hot metal that's been poured into the cavity.

Toxic fumes are produced using this method, which must be **extracted**.

Lost wax casting is a more sophisticated version of lost pattern casting using wax, which is melted before the silver or gold is cast. It's used by jewellers to produce rings and brooches.

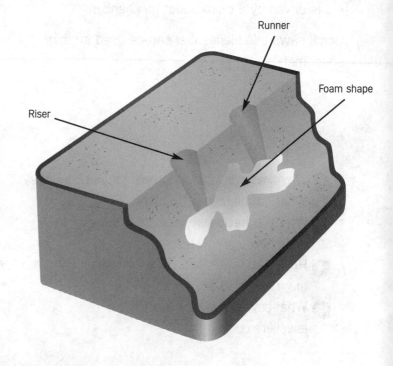

Runner
Foam shape
Riser

Casting

Industrial Die Casting

Die casting is very similar to injection moulding. It is used to manufacture large quantities of metal products. Alloys with a low melting point, e.g. pewter, aluminium, and zinc alloys can be used.

The mould is created by a spark eroding the form required into two blocks of steel. This mould is water-cooled (like a car engine) in order to control the temperature.

The following method is used:

1. The metal is heated in a **crucible** until molten.
2. A hydraulic ram pushes a quantity of the molten metal into the mould.
3. Pressure is maintained until the metal has cooled enough for the mould to be opened and the component taken out.

An Industrial Die Cast

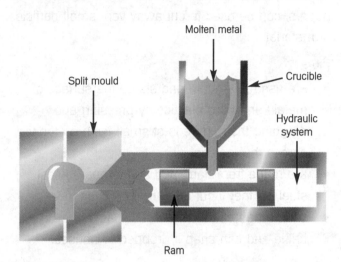

Die Casting in Schools

Die casting is often used for jewellery casting. You can die cast in school by…

- using low-melting point pewter melted with an electric paint stripper gun or a blowlamp.
- pouring the pewter into holes in easily-worked materials, e.g. cuttlefish, high-density modelling foam, MDF or acrylic

The mould can be hand cut or cut with a CNC router.

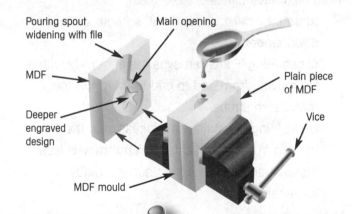

Quick Test

1. What is a cope?
2. What are the holes left by a sprue used for?
3. What are metals melted in?
4. What is lost wax casting used for?

Metal Surface Finishes

Files

Abrading tools, for example, files and abrasive papers, can be used to cut away very small particles of material.

Files…

- are used to smooth and shape the surface of metals and hard plastics by pressing and dragging the hundreds of small teeth on the file across the material
- are made from hardened and tempered cast steel, so they'll cut other metals, even steel
- should be treated with care because they're brittle and can snap if dropped or abused.

Files come in different shapes including **flat**, **half round**, **square**, **round** and **three square**. They also come in **warding** and **needle** versions in these shapes, which are very small and are used for jewellery.

Files also have different sized teeth:

- **Rough**, **bastard**, **second cut**, **smooth cut** and **dead smooth**.
- **Cross filing** – this removes metal quickly with the file moving from end to end to bring the rough material to shape.
- **Draw filing** – a finishing process with the file moving from side to side, which removes less material and leaves a finer surface ready for abrasives.

A File

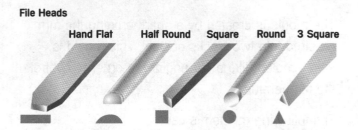

File Heads

Hand Flat   Half Round   Square   Round   3 Square

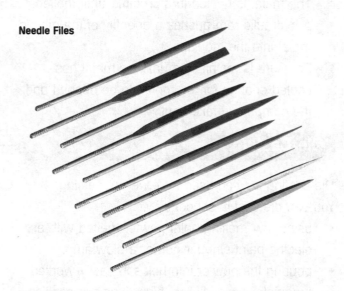

Needle Files

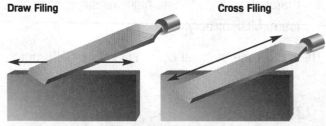

Draw Filing   Cross Filing

Abrasive Papers / Cloths and Machines

To make abrasive paper, small chips of abrasive material are glued onto a paper or cloth backing sheet. The abrasive material is usually silicon carbide or emery.

Each sheet is numbered. The lower the number, the coarser the sheet.

Emery cloth is often torn into strips and used in a two-handed manner. It's designed for use on metals although it's sometimes useful for finishing hard plastics.

Some sanding machines are designed for abrading metals and are called **linishers**. They have a continuous belt of abrasive moving across a flat plate.

Metal Surface Finishes

Polishing Metals

Metal polish is always slightly abrasive because it relies on cutting away the surface of the metal until it's very smooth. Metal polishes can be in...

- liquid form (applied with a cloth)
- a wax bar (applied to a buffing wheel).

Many products are **self-finished**. Stainless steel sheet doesn't need a surface finish because it has a very flat smooth surface, but the edge still needs to be polished.

Coatings

Polythene is the most common thermoplastic powder used for **plastic dip-coating**. It's commercially used for products such as dishwasher racks, and often used for school projects, e.g. coat hooks and tool handles.

1. Air is blown through the powder to make it behave like a liquid.
2. Metal, pre-heated to 180°C, is dipped in the fluidised powder.

3. The metal is returned to the oven where the plastic coating melts to form a smooth finish.

Powder coating is an industrial finish that's a more sophisticated version of dip-coating. The powder is sprayed onto the products which then pass through an oven. Modern powder coating...

- provides a very hard paint-like finish
- is available in all colours as well as translucent
- is very durable.

Anodising, Plating and Galvanising

Anodising is used on aluminium to provide a durable corrosion-resistant finish:

- It involves electrolysis and uses acids and electric currents, which are hazardous in school workshops.
- Colour can be added to dye the aluminium.

Plating also uses electrolysis. There are many forms, although chromium plating is the most widely recognised. The thin layer of metal on the surface provides a durable finish that is resistant to corrosion.

Galvanising involves dipping metal (usually mild steel) into a bath of molten zinc. Although it isn't very attractive, the zinc provides a very corrosion-resistant finish.

Quick Test

1. What is a file made from?
2. Why shouldn't files be dropped?
3. Name the three types of file teeth.
4. Where would it be necessary to use a 'three-square file'?
5. How should stainless steel be finished?

KEY WORDS

Make sure you understand these words before moving on!

- Files
- Flat
- Half round
- Square
- Round
- Three square
- Warding
- Needle
- Bastard
- Second cut
- Smooth cut
- Cross filing
- Draw filing
- Emery cloth
- Linishers
- Metal polish
- Plastic dip-coating
- Powder coating
- Anodising
- Plating
- Galvanising

Practice Questions

1 A non-tarnishing metal is needed for a jewellery project. Which of the following would be a possible metal for this project? Tick the correct option.

A Silver ⬭   **B** Copper ⬭

C Gold ⬭   **D** Brass ⬭

2 Which of the following statements describe valid reasons for applying a finish to a ferrous metal? Tick the correct options.

A To help protect the surface from moisture ⬭

B To stop it from being bent ⬭

C To save time when finishing ⬭

D To help protect the metal from rusting ⬭

E To improve the appearance of the surface ⬭

F To cover up the mistakes that have been made ⬭

3 Complete the chart below.

Material	Properties	What is it Used For?
Steel	a)	b)
Cast iron	c)	d)
Brass	e)	f)
Aluminium	g)	h)

4 The table contains the names of six metal finishing methods. Match descriptions **A, B, C, D, E** and **F** with the methods **1–6** in the table. Enter the appropriate number in the boxes provided.

	Wood Joints
1	Plastic dip-coating
2	No finish required
3	Anodising
4	Plating
5	Galvanising
6	Metal polish

A A rust-proof layer on a steel dustbin ⬭

B A coating for mild steel applied by dipping hot metal into a tank of plastic powder ⬭

C A thin layer of tin to seal the inside of a food can ⬭

D The surface of stainless steel ⬭

E The surface of aluminium ⬭

F A small copper box ⬭

5 What do the letters stand for in the following acronyms?

a) MIG ....

b) TIG ....

c) TPI ....

6 Which files would be suitable for the following situations?

File	Type of Teeth	Situation
a)	**b)**	Removing a lot of steel quickly from a flat steel bar
c)	**d)**	Filing the inside of a square hole in steel
e)	**f)**	Filing a concave surface (curved inwards) from a flat piece of steel
g)	**h)**	Draw filing a convex surface (curved outwards)

7 What would be a suitable steel for making chisels and drills? Tick the correct option.

A Cast iron ⬭   **B** Tool steel ⬭

C Mild steel ⬭   **D** Stainless steel ⬭

8 What would be a suitable steel for making nuts and bolts? Tick the correct option.

A Stainless steel ⬭   **B** High speed steel ⬭

C Mild steel ⬭   **D** Cast iron ⬭

9 Choose the correct words from the options given to complete the following sentences.

cross   line   dot   ruler   odd-leg callipers   hole punch   centre punch   scriber   dot punch

When marking out the centre of a hole on a piece of metal, make a .... using the

.... Make a small dent using a .... at the centre of the cross

then deepen it with a .... ready for drilling.

Plastics

Polymerisation

Plastics are the most widely used materials in commercial production.

They can be created from two main sources:
- **Natural** plastics – including materials such as amber (fossilised tree resin) and latex (a form of rubber).
- **Synthetic** plastics – by far the most common and chemically manufactured from carbon-based materials, e.g. crude oil, coal and natural gas.

Synthetic plastics are manufactured using a process known as **polymerisation**. Polymerisation occurs when **monomers** join together to form long chains of molecules called **polymers**.

Polymerisation is derived from the Greek words *poly* which means 'many' and *meros* which means 'part'. For example, Polystyrene is made up of single monomers of styrene, joined together to form a long chain.

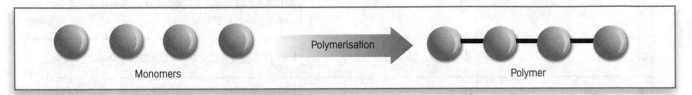

Monomers   Polymerisation   Polymer

Types of Plastic

There are two different types of plastic:
- **Thermosetting plastics** are heated and moulded into shape. They can't soften if reheated because the polymer chains become interlinked during the moulding process. Individual monomers are joined together to form a massive polymer.

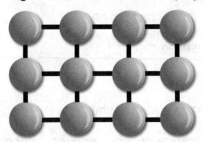

- **Thermoplastics** soften when they're heated and can be shaped when hot. The plastic will harden when it's cooled, but can be reshaped if heated up again. This is called **post forming** from the Latin *post* meaning after, e.g. p.m. (afternoon).

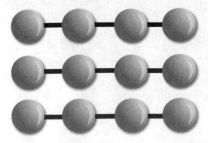

There are several factors to consider when choosing a plastic for a specific purpose:
- **Colour** – different plastics differ greatly in colour.
- **Texture** – different plastics have varied surface and cell textures.
- **Workability** – some plastics are much easier to work with than others.
- **Structural strength** – different plastics vary from weak to very strong.
- Whether the plastic can be **recycled** or not.
- Whether to use a **thermoplastic** or a **thermosetting plastic**.

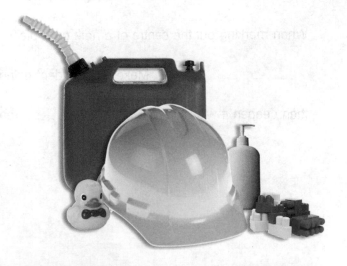

Thermoplastics

Name	Description	Uses
Polythene (high density) HDPE	• Stiff, strong plastic • Softens at 120-130°C	• Pipes and bowls • Milk crates • Buckets
Polythene (low density) LDPE	• Weaker, softer and more flexible than HDPE • Softens at 85°C	• Packaging • Film • Carrier bags • Toys • 'Squeezy' detergent bottles
Polypropylene (PP)	• High impact strength • Softens at 150°C • Can be flexed many times without breaking	• Bottle crates and boxes • Medical equipment and syringes • Food containers • Nets • Storage
High Impact Polystyrene (HIPS)	• Light but strong • Widely available in sheets • Softens at about 95°C	• Vacuum forming • Very common for school project work, e.g. outer casings on electronic products, packaging
Nylon (Polyamide)	• Hard material • Good resistance to wear and tear • Solid nylon has low friction qualities and a high melting point	• Curtain rail fittings • Combs • Hinges • Bearings • Clothes • Gear wheels
Rigid PVC (Polyvinyl chloride)	• Stiff and hard wearing • Plasticiser can be added to create a softer, more rubbery material	• Air and water pipes • Chemical tanks • Shoe soles • Shrink and blister packaging • Floor and wall coverings
Acrylic (Polymethyl-methacrylate)	• Trade name is Perspex • Glass-like transparency or opaque. Can be coloured with pigments • Hard wearing, but will shatter if treated roughly	• Display signs • Baths • Roof lights • Machine guards

Plastics

Name	Description	Uses
Melamine formaldehyde (Methanal, MF)	• Heat resistant polymer	• Tableware • Electrical installations • Synthetic resin paints • Decorative laminates • Worktops
Epoxy resin (Epoxide, ER)	• A resin and a hardener mixed to produce a casting	• Castings • Printed circuit boards (PCBs) • Surface coating • Araldite Glue™
Polyester resin (PR)	• A resin and hardener mixed together • Polymerises at room temperature • Often reinforced with glass fibre	• Laminated to form glass reinforced plastic (GRP) castings • Encapsulations • Car bodies • Boats
Phenol formaldehyde	• Also known as Phenol Methanal (PF) or Bakelite • Hard, brittle plastic • Dark colour with glossy finish • Heat resistant	• Dark coloured electrical fittings and parts for domestic appliances • Kettle / iron / saucepan handles
Urea formaldehyde	• A colourless polymer • Coloured with artificial pigments to produce a wide range of different colours	• Door and cupboard handles • Electrical switches • Electrical fittings

Measuring Lengths and Angles

A **steel rule** is used to…
- measure lengths when marking on plastics
- set a compass.

A **tri square** is used to…
- mark lines exactly at 90° from the edge of the material
- check that the material has square corners.

You need to have one accurate straight edge on the material before you can use this tool.

A **sliding bevel** is an adjustable angle marker that can be set to any angle.

Sliding Bevel

Marking the Surface

When marking out plastics, it's sensible to leave the protective covering on the surface for as long as possible. You can mark out on this covering.

You can use the following for marking on plastics:
- **Pencil** – for marking on the paper covering. It's soft and will not scratch the surface.
- **Fine line marker** – marks the surface of the plastic covering or directly on the plastic surface. You can wipe the ink off with a solvent.
- **Compass** – used to mark out circles or arcs. Put a piece of masking tape at the centre of the circle to prevent the point digging into the material.
- **Centre punch** – used for marking the centre of holes for drilling into plastic. The punch leaves a

small indentation which prevents the drill bit from wandering across the surface.
- **Card stencils** – used for marking out curved shapes onto any material. Symmetrical shapes are best done by folding the template in half and cutting both sides together.

Stencil

Centre Punch

Compass

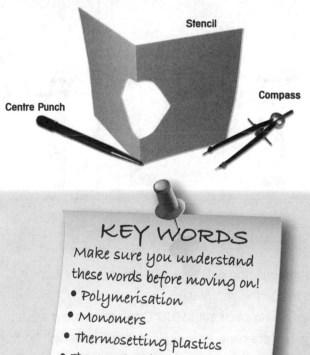

Quick Test

1. Name the two main types of plastic.
2. What are the two main characteristics of each type of plastic?
3. Name a plastic that would be used for making electrical fittings and say why it's the best material.
4. Which three tools are used to mark a line around the end of a square piece of acrylic?
5. Why would you use a fine-line marker when marking out a line on plastic?

KEY WORDS

Make sure you understand these words before moving on!
- Polymerisation
- Monomers
- Thermosetting plastics
- Thermoplastics
- Post forming

Moulding Plastics

Line Bending

Acrylic sheets are often bent using a **line bender** or **strip heater**:

1. The plastic sheet is heated along the line of the intended fold.
2. The softened area is then bent to shape using a **bending jig** for accuracy.

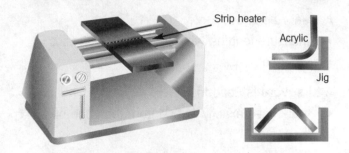

Strip heater

Acrylic

Jig

Vacuum Forming

Vacuum forming uses thermoplastic materials in the form of sheets that can measure up to 1.5m x 1.8m. The most popular material is high-impact polystyrene (HIPS), which is cheap and easy to form.

Heated plastic is 'sucked' onto the **former**, which forms the required shape. A former can be a wooden **mould** around which the softened plastic will be held by the vacuum until it has cooled.

1. The plastic is heated and the mould moves close to it. Air is 'sucked out' to form a vacuum.

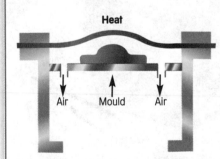

Heat

Air   Mould   Air

2. This causes the hot plastic to be sucked onto the mould. As the temperature of the plastic falls, a rigid impression of the mould is formed.

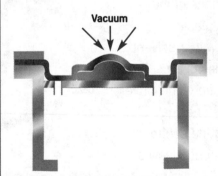

Vacuum

3. The vacuum pump is turned off allowing air to enter. The former is lowered, separating it from the final product.

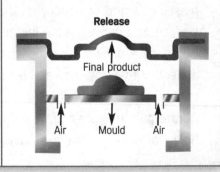

Release

Final product

Air   Mould   Air

Yoke Moulding

Yoke moulding only requires one accurate former to go inside the soft plastic:

1. Use an oven set to 110°C to soften the material.
2. The pressure from a close-fitting yoke will hold the plastic until it has hardened.

This technique is used for making half-shells, which can be solvent welded together. Foamed PVC is a good material to use because it has a long working time once hot.

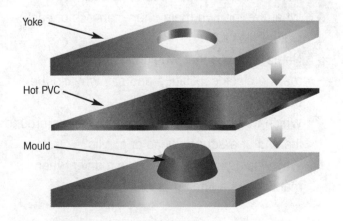

Yoke

Hot PVC

Mould

Injection Moulding

Typical materials used in **injection moulding** are polythene, polystyrene, polypropylene and nylon.

You need to know the following steps:

1. Plastic powder or granules are fed from the hopper into a hollow steel barrel.
2. The heaters melt the plastic as the screw moves it along towards the mould.
3. Once enough melted plastic has collected, the screw forces the plastic into the mould.
4. Pressure is maintained on the mould, until it has cooled enough to be opened.

Injection moulding can be done in school by using

- a hot glue gun
- a simple mould made from two pieces of steel with a cavity between them.

The hot glue is squeezed in through the hole and left to cool.

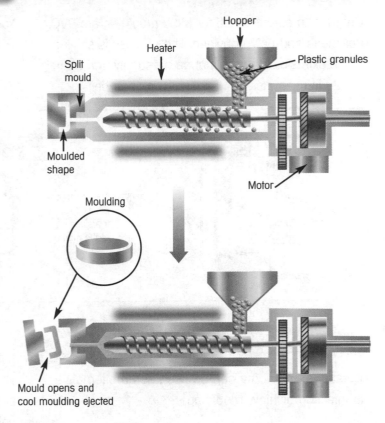

Hopper

Heater

Split mould

Plastic granules

Moulded shape

Motor

Moulding

Mould opens and cool moulding ejected

Extrusion

Typical materials used in **extrusion** are polythene, PVC and nylon.

You need to know the following steps:

1. Plastic granules are fed into the hopper by the rotating screw.
2. The plastic granules are heated as they're fed through.
3. The softened plastic is forced through a die in a continuous stream, to create long tube or sectional extrusions (different from injection moulding).
4. The extrusions are then passed through a cooling chamber and cut to the required length.

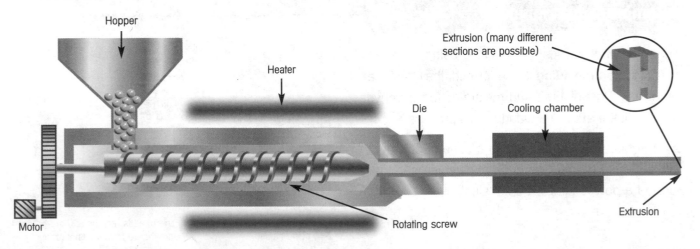

Hopper

Heater

Extrusion (many different sections are possible)

Die

Cooling chamber

Rotating screw

Motor

Extrusion

Moulding Plastics

Blow Moulding

Common materials used for **blow moulding** are PVC, polythene and polypropylene. The process is similar to the extrusion process, but an air supply and a **split mould** is used instead of the cooling chamber.

You need to know the following steps:

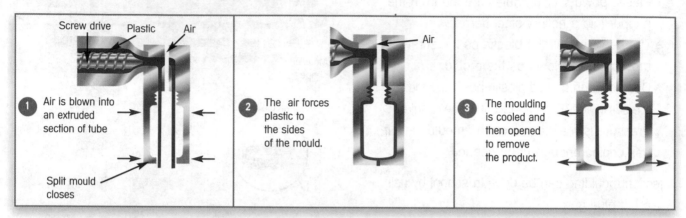

1. Air is blown into an extruded section of tube. Split mould closes

2. The air forces plastic to the sides of the mould.

3. The moulding is cooled and then opened to remove the product.

Rotational Moulding

Rotational moulding can be used as an alternative to injection or blow moulding.

A rotational moulding machine is rotated continuously with heated thermoplastic powder inside.

This method is used to make footballs, traffic cones and storage tanks. The mouldings are made from polythene (PE), which has fire retardant qualities.

You need to know the following steps:

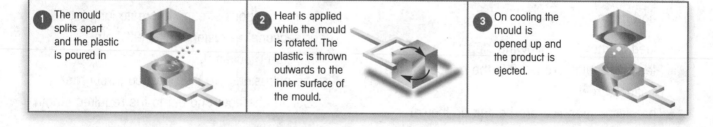

1. The mould splits apart and the plastic is poured in

2. Heat is applied while the mould is rotated. The plastic is thrown outwards to the inner surface of the mould.

3. On cooling the mould is opened up and the product is ejected.

Compression Moulding

Thermosetting plastics are moulded using **compression moulding**. Once formed, they can't be re-formed. Phenol, urea and melamine formaldehyde are plastics that are moulded in this process.

A large force is used to squash a cube of polymer into a heated mould. The cube of polymer is in the form of a powder, known as a 'slug'.

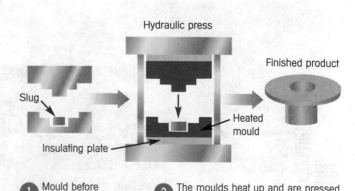

1. Mould before being heated

2. The moulds heat up and are pressed together to form the final product

Cutting Plastics

Sawing

When sawing plastics...
- the blades are held in tension within a frame
- fine blades should be fitted with more than 30TPI
- it's an easy process to change the blade when it becomes worn or damaged
- leave the protective film on the plastic while cutting
- use a vice or a G Cramp to hold the plastic
- protect the plastic from the jaws with a soft insert.

There are several different saws:
- A coping saw (sawing curves).
- A hacksaw (sawing straight lines).
- A junior hacksaw (sawing straight lines very accurately)
- A power bandsaw, which rotates a continuous strip of saw blade. Only fine blades should be used for plastics.

- A power scroll saw, which uses a reciprocating motion with the blade held in tension. The blade moves up and down through a table that can be angled. Fine blades are available for plastics.

Tip: To prevent the plastic from sticking to the blade when being sawn with a scroll saw, put a piece of masking tape over the surface to be cut.

Scroll Saw

Drilling and Hand Planing

Twist drills or jobbers drills are...
- used for drilling smaller diameter holes in plastics
- unsuitable for larger diameters as they leave a ragged edge to the hole (a hole saw can be used for larger diameter holes).

Some planes are specially adapted for plastics, e.g. using a block plane and sometimes a surform.

Quick Test

1. What is a bending jig?
2. How does a vacuum former work?
3. What do the letters PVC stand for?
4. By which process are PVC drink bottles made?
5. Which process is used for moulding thermosetting plastics?

KEY WORDS

Make sure you understand these words before moving on!
- Line bender
- Strip heater
- Bending jig
- Vacuum forming
- Former
- Mould
- Yoke moulding
- Injection moulding
- Extrusion
- Blow moulding
- Split mould
- Rotational moulding
- Compression moulding

Joining and Finishing Plastics

Nuts and Bolts

Mechanical joining methods are used to join plastics to themselves or to other materials.

- Some **bolts** are called machine screws and have the thread over the entire length.
- **Nuts** must match the same thread as the bolt and they come in a variety of types.
- A **washer** is usually used under the nut to protect the plastic surface. It might be a plain ring, or a sprung washer, which is used to stop the nut from vibrating loose.

Female threads are sometimes cut into one piece of the plastic and a machine screw is then used to hold plastic components together. **Plastic** threads are usually injection moulded rather than being cut with a tap and die.

Metric sized threads are almost standard in schools (M3-M12 are the most popular). They're available in many lengths, normally 20-100mm.

Adhesives and Solvent Cement

Some plastics can be thermally bonded by heating, but adhesives and solvents are often used:

- **Hot melt glue** – used for quick modelling, but not in final products.
- **Epoxy resin** (Araldite™) – a very versatile but expensive adhesive which sticks most clean dry materials. Equal amounts of resin and hardener are mixed together. It sets chemically to become very hard.

There are several types of **solvent cement** available. The most common is Dichloromethane, which dissolves the surface of hard plastics, e.g. acrylic and high impact polystyrene.

Ventilation is essential as very dangerous fumes are given off.

Self-finishing

Many plastics can be **self-finished**, e.g. injection moulded products. The mould is highly polished or textured and this ensures that the same surface is transferred onto each product.

The surfaces of sheet plastics don't usually need to be finished unless they've been damaged. This is why they have a layer of paper or polythene on the outside.

The cut edges of plastic materials need to be filed, abraded and then polished to a high shine.

- A smooth file can be drawn along the edge of the plastic (draw filing) until only fine lines can be seen running along the edge.

- Different grades of **wet and dry paper** are then wetted with water and used to remove the lines until the edge is smooth.

Polishing Plastics and New Materials

Polishing Plastics

Hard plastics, e.g. acrylic, are often polished on their cut edges. Polishing can also be used to remove fine scratches by using a...

- metal polish applied by hand with a cloth
- compound applied using a **buffing wheel**.

Polishing compounds, e.g. Vonax, can achieve a high gloss surface. But, it's easy to overheat the edge of the plastic (by pressing too hard onto the buffing wheel) and this can permanently damage the surface.

New Materials

There is a growing group of new materials that have different properties from traditional materials.

- **Smart materials** react to outside stimuli and change their properties as a result, for example, they become different shapes or colours.
- **Composites** are made by joining two or more materials to gain a new material that has desirable properties, for example, greater elasticity and strength.

Reactolite Glasses

Changes to dark in sunlight

Smart Alloys

A smart alloy, or **shape memory alloy (SMA)**, is a material that can remember its original shape, e.g. nickel-titanium, copper-zinc-aluminium and copper-aluminium-nickel:

- When the alloy is bent or twisted, it keeps its new shape until it's heated.
- When the temperature is raised to a certain level the alloy returns to its original shape.

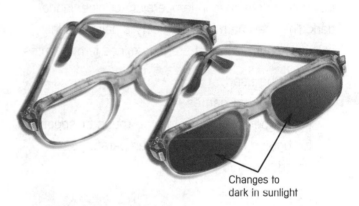

Original shape → **Force** → Deformed shape → **Heat** → Original shape

SMAs can be made in wire form, i.e. **smart wire**:

- When a small electric current is passed through the wire it shrinks in length.
- When the current stops it returns to its original size.

One application of this is in operating lightweight mechanisms. Smart springs are also available.

Smart Wire   **Smart Springs**

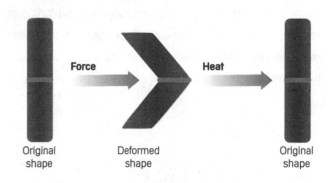

SMART WIRE

Smart and Composite Materials

Smart Steel and Plastics

Smart steel components can...

- become magnetic under high stress (used to monitor conditions in machines)
- absorb impact (being developed to improve car passenger safety).

Polymorph is a tough polymer (plastic) which softens and becomes easy to mould at only 62°C. It can be softened with hot water or a hairdryer, moulded into shape by hand and then hardened again when cool.

It can be used for prototyping tool handles or other organic shapes. Conductive polymers are plastics that can conduct electricity.

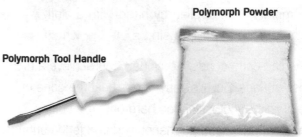

Polymorph Powder

Polymorph Tool Handle

Smart Colours

Smart colours are a range of **thermochromic** pigments that react to changes in temperature or glow in the dark. They can be mixed with acrylic paint and the colour change occurs at pre-determined temperatures.

They can be used...
- for designing games
- for temperature indicators on mugs or spoons for babies to indicate safe temperatures
- as additions to jewellery.

Thermochromic

Other Smart Materials

Smart grease is a very sticky and viscous gel that can be used to control the movement of mechanisms. For example, on a rubber band-driven toy it can regulate the speed at which the potential energy is released.

Piezo-electric actuators are...
- quartz crystals that have become electrically charged due to being deformed under mechanical stress. (This situation can be reversed and the same materials undergo dimensional change under the influence of an electric field.)
- used as gas lighters on cookers.

Composites

Glass reinforced polyester (or plastic) **(GRP)**...
- is polyester resin that has been reinforced by the addition of strands of spun glass fibres
- makes a light, hard-wearing material that is durable and resistant to corrosion
- is used in car body building and boat hulls.

Reinforced concrete is...
- cement mixed with water, sand and aggregate and reinforced by the addition of steel bars
- a dense material that is very strong under compression (concrete), but also flexible and able to take tensile forces (steel)
- used for the construction of buildings and bridges.

Smart and Composite Materials

Other Composites

Carbon fibres are spun into a cord, which is laminated with plastics to give exceptional strength and resistance to strain. They are used in camping and sports equipment.

Threads of **Kevlar** retain much of their strength at high temperatures and under stress. Kevlar is used in protective equipment, for example, cut-resistant gloves, helmets, vehicle protection and body armour.

Kevlar threads…
- help the seams and edges stay sealed
- keep out flames or sharp points that would burn or cut the wearer.

Tufnol is a light-weight, high-pressure laminate, which has good strength and electrical insulating properties. It doesn't corrode and has excellent weather resistance and chemical resistance to most oils. It's used in situations requiring very long life in outdoor or marine environments.

A **bi-metallic strip** is a composite of two metals with different rates of expansion. When the temperature changes, the metal with the greater rate of expansion bends round the other. They're used for simple temperature sensing devices, e.g. thermostats in heating systems.

Nano Materials

Nano materials…
- are based on very small nanoparticles with diameters of below 50 nm (the size of molecules)
- have useful properties as lubricants, flame retardants, sealants, binders and adhesives
- are strong UV light absorbers, so they can be used as a protective pigment in paints
- are used in self cleaning glass
- are small enough to enter the bloodstream, so they can be used to deliver drugs round the body.

Nanotechnology can be used to increase the strength of polymers and make surfaces harder wearing. It can also be used to help create…
- clothing that repels dirt, stains and body odours and can 'self clean' with a cup of water
- garments that can sense, react and absorb an impact or collision and so protect your body
- sports clothing that can measure your individual fitness levels and create individual training programmes based on your body's feedback.

Quick Test

1. What is used to shape soft plastic?
2. What could be used to 'glue' plastics together?
3. Where would a photochromic material be useful?
4. What materials make up reinforced concrete?
5. Where would a bi-metallic strip be used and what does it do?

KEY WORDS
Make sure you understand these words before moving on!
- Solvent cement
- Self-finished
- Wet and dry paper
- Buffing wheel
- Smart materials
- Composites
- Shape memory alloy
- Smart wire
- Thermochromic
- Piezo-electric actuators
- Glass Reinforced Polyester (GRP)
- Reinforced concrete
- Carbon fibres
- Kevlar
- Tufnol
- Bi-metallic strips
- Nano materials

Practice Questions

1 Circle the correct options in the following sentences.

a) A **thermosetting** / **thermoplastic** material will not be affected by re-heating.

b) A **thermosetting** / **thermoplastic** material is ideal for post forming.

2 A non-flammable plastic is needed for an electric switch. What would be one possible plastic? Tick the correct option.

A High impact polystyrene ◯   **B** Acrylic ◯

C Urea formaldehyde ◯   **D** Epoxy resin ◯

3 Which of the following statements is **not** a valid reason for choosing a plastic for vacuum forming? Tick the correct option.

A It should be recyclable ◯

B It should have a wide range of colours ◯

C Once formed it shouldn't change shape unless it is re-heated ◯

D It should be easy to bend when cold ◯

4 Give two properties of phenol formaldehyde.

a) ....   **b)** ....

5 Give two uses of polythene.

a) ....   **b)** ....

6 Give two properties of nylon.

a) ....   **b)** ....

7 What do the letters stand for in the following acronyms?

a) PVC ....   **b)** HIPS ....

c) HDPE ....

8 Describe a) the tools and the equipment, and b) the process used to achieve a high quality finish for the sawn edge of a 5mm piece of acrylic.

a) ....

....

b) ....

....

9 The table contains the names of six plastic forming methods. Match descriptions **A, B, C, D, E** and **F** with the methods **1–6** in the table. Enter the appropriate number in the boxes provided.

Plastic Forming Method			
1	Vacuum forming	**4**	Extrusion
2	Yoke moulding	**5**	Compression moulding
3	Injection moulding	**6**	Rotational moulding

A A sheet of foamed PVC is heated in an oven and then pressed between a former and a shaped frame.

B A thermosetting plastic powder slug is pressed into shape and heated in a mould.

C Polythene chips are placed in the mould which is then sealed, heated and turned round until the plastic has coated the inside. When cool the mould opens and the finished item is ejected.

D A sheet of polystyrene is held in place, heated and then shaped to a former by sucking out the air between the plastic and the former.

E Nylon granules are melted and then forced through a die in a continuous stream to form a pipe.

F Polypropylene granules are fed into a heated chamber and then pushed into a sealed mould where the moulding takes place

10 What composite material would be suitable to use for making foundations, walls and bridges? Tick the correct option.

A Brick

B Stainless steel

C Laminated wood

D Reinforced concrete

11 Choose the correct words from the options given to complete the following sentences.

shape   mould   squeeze   inject   female   male   threader   tap

To make a bolt from polystyrene, it's necessary to use a _____ and

_____ hot plastic into the cavity. The matching _____ thread can

be cut by using the correct _____.

Milling, Routing and Turning

Milling and Routing

Wood, plastics and metal can all be shaped using a revolving, multi-toothed cutter, which moves over the material. This technique is known as...

- **milling** if you're shaping metals and plastics
- **routing** if you're shaping timber.

Shapes can be cut manually using a powered router. The router can follow a template, used with a guide to cut slots or to shape the edge of a timber board.

A hand-held router has a guide attached for cutting slots parallel to the edge of a board. This has a variety of cutters that can make different profiles on the material. The cutter rotates clockwise into the material.

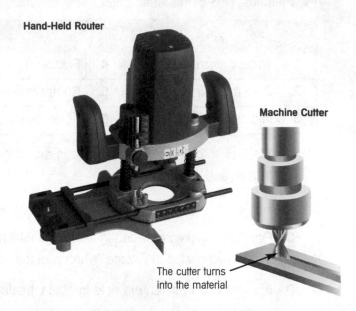

Hand-Held Router

Machine Cutter

The cutter turns into the material

CNC Milling

Traditional milling machines can be controlled by moving each axis manually.

By moving each axis with a stepper motor, very accurate movements can be controlled using computer numerical control (CNC). This is one of the most common forms of computer aided manufacture (CAM).

CNC routers are very common in the furniture industry.

Centre Lathes

Turning involves rotating the work against a blade.

Turning metals and plastics on a **centre lathe** involves holding the work (usually in a chuck) and rotating the work towards the cutter. The cutter can be moved left and right (x), forwards and backwards (in and out (y)).

Five different types of tool are used – **roughing tools**, **knife tools** (**left** and **right handed**), **round nosed tools**, **parting tools** and **boring tools**.

The **tailstock** can be used to support long pieces of material or fitted with a drill chuck for drilling holes into the end of the material.

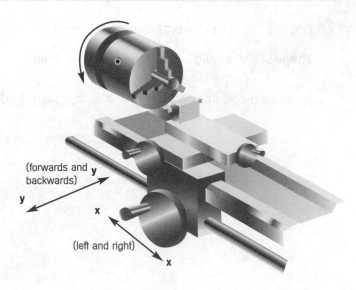

(forwards and backwards) **y**

y

x

(left and right)

x

Wood-turning Lathe

On a **wood-turning lathe**, the tool is rested on a support and is guided by hand. The work can be held between centres or screwed onto a faceplate.

Four different types of tool are used:

- **skew chisels**
- **gouges**
- **parting chisels**
- **scrapers**.

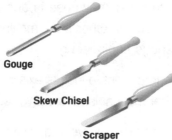

Gouge

Skew Chisel

Scraper

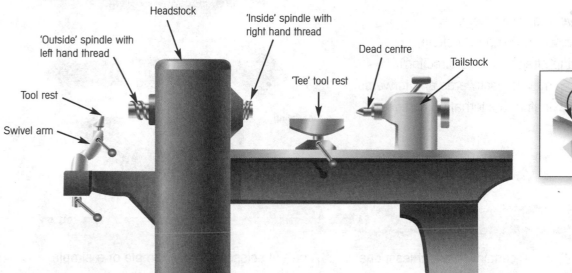

'Outside' spindle with left hand thread

Headstock

'Inside' spindle with right hand thread

Dead centre

Tailstock

Tool rest

'Tee' tool rest

Swivel arm

Turning

CNC Turning

The movements of both the work and the cutting tools can be controlled on centre lathes using **stepper motors**. This allows the lathe to be numerically controlled.

CNC lathes are particularly useful for turning quantities of identical pieces.

Quick Test

1. What do the letters CNC stand for?
2. Give two uses for a tailstock.
3. Which way does a router cutter turn?
4. What type of motor is used to control a CNC milling machine?

KEY WORDS

Make sure you understand these words before moving on!

- Milling
- Routing
- Centre lathe
- Roughing tool
- Knife tools (left and right handed)
- Round nosed tools
- Parting tools
- Boring tools
- Wood turning lathe
- Skew chisels
- Gouges
- Parting chisels
- Scrapers
- Stepper motors

Levers

A mechanism creates movement within a product. A designer may need to apply a mechanism to any product, whether it's a moving toy or a piece of serious engineering.

It's important to look at existing products and work out what mechanisms are involved in making them move.

There are four types of movement:

- **Rotating** or **rotary** (turning in a circle).
- **Linear** (moving straight in one direction).
- **Reciprocating** (moving backwards and forwards).
- **Oscillating** (swinging in alternate directions).

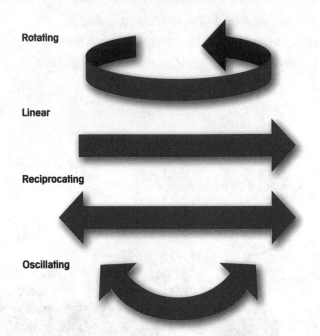

Rotating

Linear

Reciprocating

Oscillating

Lever 1 – Basic Principles

All machines will almost certainly have at least one **lever.** A lever is a simple device, consisting of a rigid bar which pivots about a fixed point (the **fulcrum**).

A pair of scissors is an example of a simple lever system.

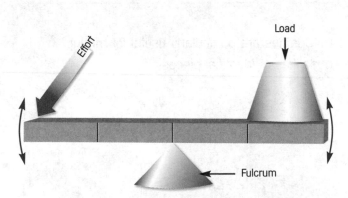

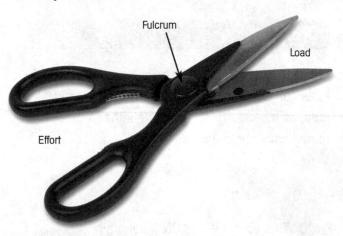

- A 'load' is applied at one end of a 'rigid bar'.
- The bar is placed centrally on top of the fulcrum.
- At the other end of the bar a force is applied (the effort).
- This results in a single 'lever' movement about the fulcrum.

- The effort is applied by the hands at one end.
- The load is the resistance against the cutting edge.
- The fulcrum is the screw that holds the two halves together and allows for movement.

First Class Levers

Levers can be force multipliers, e.g. using a crowbar. By altering the position of the fulcrum, the effort can be multiplied and a larger load can be lifted by a **first class lever**. This is called the **mechanical advantage**.

The fulcrum is between the **load** and the **effort**. The effort needed is less than the load, because the load is nearer to the fulcrum.

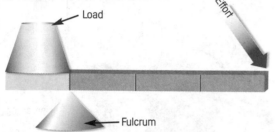

The leverage of the blue sections on the rigid bar, are at a ratio of 1:3, so an effort of 1 could move a load of 3, but the effort end will have to move 3 times further than the load end.

Second Class Levers

A **second class lever** is found where the load is applied between the effort and the fulcrum. The effort needed is less than the load, because the load is nearer to the fulcrum.

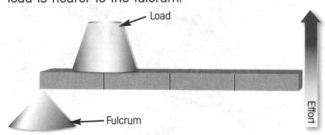

Nutcrackers and wheelbarrows are examples of how a lever can be used to act as a force multiplier.

Third Class Levers

A **third class lever** is where the effort is applied between the load and the fulcrum. The effort needed is greater than the load, but this time the amount of movement of the load is multiplied.

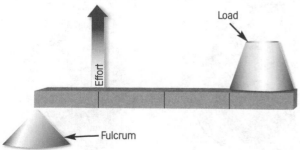

The elbow is the fulcrum. The effort is provided by the biceps muscle, which attaches to the forearm just below the elbow.

A relatively small movement of the biceps results in a relatively large movement of the end of the lower arm, but the effort needs to be greater than the load.

Cranks and Cams

General Mechanical Movement

Cranks and cams are relatively simple devices which convert **rotary motion** to **reciprocating motion** (or vice versa).

Cranks

Cranks can convert reciprocating motion to **rotary** motion, for example in the crankshaft in a car engine.

When arranged differently, cranks can also convert **rotary** motion to **linear** or reciprocating motion, for example, in…

- driving a pump
- tricycles
- children's pedal cars

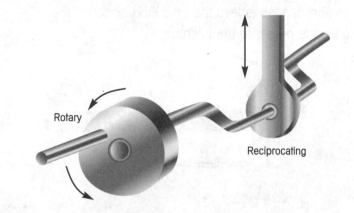

Rotary

Reciprocating

Cams

A **cam** is a device that usually converts **rotary** motion into **reciprocating** (up and down) motion.

For example, in a rotary cam the reciprocating motion is in the cam follower. This motion can be varied by using different shaped cams.

A **follower** is a rod that moves up and down. This will have an object of your choice on top.

A **guide** holds the follower in place.

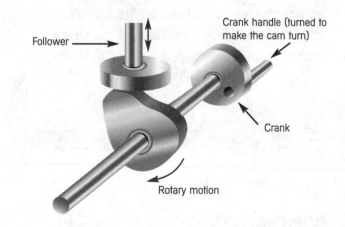

Follower

Crank handle (turned to make the cam turn)

Crank

Rotary motion

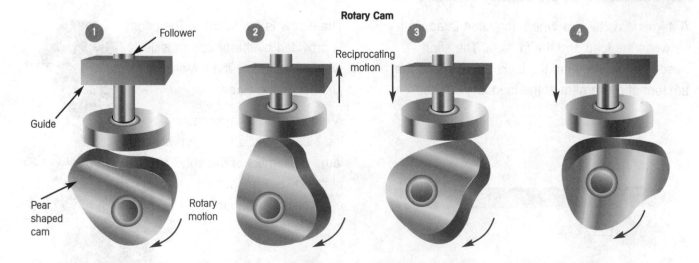

Rotary Cam

1   Follower

Guide

Pear shaped cam

Rotary motion

2   Reciprocating motion

3

4

Springs and Linkages

Springs

There are many different types of spring that are used to resist different forces. They can be placed into four broad groups:

- Springs that resist **extension**.
- Springs that resist **compression**.
- Springs that resist **radial movement**.
- Springs that resist **twisting**.

All springs resist force by trying to return to their original shape or position.

Linkages

Linkages can sometimes act as levers, but they mostly transfer one **mechanical** motion to another. Linkages are often used to connect cams to cranks or cams to levers or vice versa.

An example is a metal tool box which opens to reveal different levels of trays.

Three simple linkage systems are moving wings, tongs and push-pull.

Moving Wings

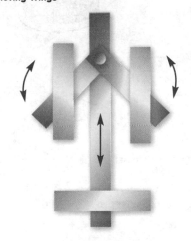

Push–Pull

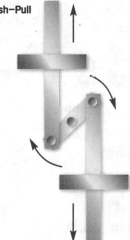

Tongs

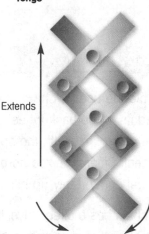

Extends

Quick Test

1. List the four main types of movement.
2. What is the proper name for a pivot?
3. What kind of motion do cranks and cams convert?
4. List the four main types of spring.

KEY WORDS
Make sure you understand these words before moving on!
- Rotating
- Rotary
- Linear
- Reciprocating
- Oscillating
- Fulcrum
- First class lever
- Mechanical advantage
- Load
- Effort
- Second class lever
- Third class lever
- Cranks
- Cams
- Follower
- Guide
- Extension
- Compression
- Radial movement
- Twisting
- Linkages

Gears and Pulleys

Gears

Gears are linkages for transferring motion. Gear wheels have teeth that mesh with the teeth of other gears.

Gears are used as force multipliers or reducers. For example, gears reduce the applied force needed to drive a car uphill. The small pinion moves the big pinion (with twice as many teeth), so it rotates at half the speed (but with twice the force).

A **rack and pinion** linkage is used in cars to convert the rotary motion of the steering wheel into a lateral movement of the wheels.

A **worm and worm wheel** changes motion through 90°. The worm is usually the driver. This creates a large reduction in speed and a high torque, for example, hand-held food mixers.

Bevel gears also change motion through 90°. If the gears are different sizes there will be a change in speed as well. For example, hand drills.

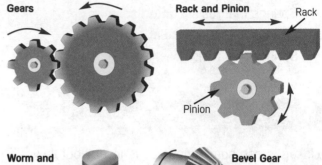

Gears

Rack and Pinion

Rack

Pinion

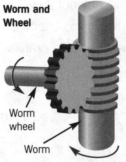

Worm and Wheel

Worm wheel

Worm

Bevel Gear

Chain and Sprocket

A bicycle uses a **chain and sprocket** to connect the pedals to the back wheel. As the pedal is pushed, the chain links with the sprocket and turns the wheel. Different sized sprockets make it possible to pedal comfortably up hill.

A chain has a direct drive. It doesn't stretch or slip, but it's difficult to change and has to be lubricated with oil.

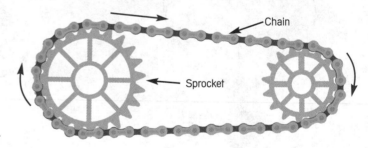

Chain

Sprocket

Pulleys

A **pulley** is a grooved wheel with a belt running in the grove. A twist in the belt makes the second pulley turn in the opposite direction.

A belt stretches, allowing it to absorb shock and be changed easily. But, a belt can also slip. They are used to…

- control how fast something turns, e.g. CD players and washing machines
- make lifting easier, e.g. cranes.

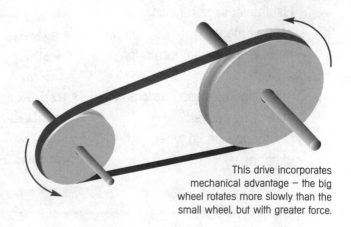

This drive incorporates mechanical advantage – the big wheel rotates more slowly than the small wheel, but with greater force.

Electrical Components

Components

Batteries provide the power source for most portable electronic circuits:
- They are compact and safe.
- Some are rechargeable (more environmentally friendly).

Wind up power sources are now available, which are very environmentally friendly.

Switches make or break a circuit. There are different kinds and they can have more than one function. For example, a Double Pole Double Throw (DPDT) switch can control two circuits at the same time.

Circuit boards don't use wiring – the components are soldered onto the board.

Lamps can be connected into a circuit as indicators or to provide light. They're fragile and can break easily.

Light emitting diodes (LEDs) can be used instead of bulbs. They use less power and are much tougher. The correct polarity must be used. High power LEDs are now used for long-life torches.

A direct current (DC) motor can be used to provide movement for a mechanism. The direction of rotation is altered by changing over the connections to the battery.

Cell	Battery	Switch	DPDT switch	Bulb	LED	Motor

Assembly

Components are held in place on a circuit board by soldering. To protect the parts from heat, a pair of pliers should be used as a heat sink by holding the 'leg' of the component while soldering.

Integrated circuits are small electronic packages that are used to perform specific operations. They are commonly used in schools as timers and counters. An integrated circuit is mounted into a holder, which has been soldered to a circuit board. These circuits are very efficient and cheap.

Quick Test

1. What are two advantages of using a chain and sprocket rather than a pulley system for a bike?
2. What are two advantages of using a pulley system rather than a chain and sprocket for a lawnmower?
3. Why are LED lights better for the environment than bulbs?

KEY WORDS

Make sure you understand these words before moving on!
- Gears
- Rack and pinion
- Worm and worm wheel
- Bevel gears
- Chain and sprocket
- Pulley
- Batteries
- Switches
- Circuit boards
- Lamps
- Light emitting diode (LEDs)
- DC motor
- Heat sink

Practice Questions

1 Circle the correct options in the following sentences.

a) A **reciprocating** / **oscillating** movement goes backward and forwards in an arc.

b) A **reciprocating** / **oscillating** movement goes backward and forwards in a straight line.

2 What is one possible mechanism needed to convert reciprocating motion to rotary motion? Tick the correct option.

A Pinion ◯   B Crank ◯

C First order lever ◯   D Cam ◯

3 What is one possible mechanism needed to convert rotary motion to reciprocating motion? Tick the correct option.

A Pinion ◯   B Fulcrum ◯

C First order lever ◯   D Cam ◯

4 **a)** Name three commonly used metal-cutting lathe tools.

i) ....   ii) ....

iii) ....

b) Name three commonly used wood-cutting lathe tools.

i) ....   ii) ....

iii) ....

5 Sketch a pulley system and explain what it is used for.

6 What do the following terms stand for with reference to a **cam system**?

a) Lift ....   **b)** Guide ....

c) Follower ....

7 The table contains the names of three lever systems. Match descriptions **A**, **B** and **C** with the methods **1–3** in the table. Enter the appropriate number in the boxes provided.

	Levers
1	First class lever
2	Second class lever
3	Third class lever

A The effort applied between the load and the fulcrum. The effort needed is greater than the load, but the amount of movement of the load is multiplied. ◯

B The load applied between the effort and the fulcrum. The effort needed is less than the load, because the load is nearer to the fulcrum. ◯

C The fulcrum is between the load and the effort. The effort needed is less than the load, because the load is nearer to the fulcrum. ◯

8 Fill in the table to show which mechanism would allow the movement of the output shaft to give the output in the following situations.

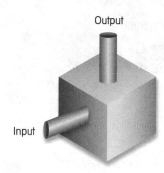

Output

Input

Input	Mechanism, Name and Sketch	Output
Rotary motion clockwise	**a)**	Rotary motion anticlockwise
Rotary motion clockwise	**b)**	Reciprocating motion
Reciprocating motion	**c)**	Oscillating motion

9 A component is required for a panel to be illuminated from behind using minimal power. Which of the following components is needed for the job? Tick the correct option.

A 240V bulb ◯   **B** Light emitting diode ◯

C 1.5V bulb ◯   **D** 24V bulb ◯

Answers

Quick Test Answers
Page 9

1. **Accept any three of the following:** Everyone can understand the drawings; It's easy to understand all the views; The views are accurate with no distortion; Measurements can be taken from the drawings; All the information for making is in the same place.

2.

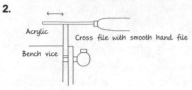

3.

4. **Accept any three of the following:** The design idea can be viewed from all sides; You get a better idea of scale; Some construction problems can be identified; The client and other users can be consulted; Modifications can be made easily.

5. **Accept any three of the following:** Sheet cardboard; Corrugated cardboard; Corrugated plastic sheet; Polystyrene sheet; Straws; Polymorph; Styrofoam.

Page 11

1. Computer Aided Design
2. CNC Milling machine / CNC Lathe Laser Cutter
3. **Accept any four of the following:** Accurate drawings can be produced; Designs can be produced quickly by skilled operators; Information can be shared by e-mail; Virtual reality images can be easily evaluated; The size or shape of a design can be quickly changed; Copies can be printed out for presentations to client; Information can be easily stored; Copies can be backed-up for security; Designing can take place in different locations; The designs can be prototyped easily using CNC machines.

4. **Accept any four of the following:** High speed of production; High quality edges which need minimal surface finishing; Multiple items will be identical; Very accurate details can be cut; Changes to the programme can be made easily.

5.

Answers to Practice Questions
Pages 12–13

1. A
2. **a)** 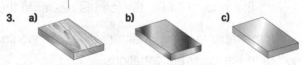 **b)** Third angle orthographic projection.

3. **a)** **b)** **c)**

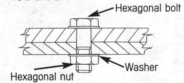

4. manufacturer; orthographic; instructions; prototype
5. B, D and E
6. A, B and D
7.

- Hexagonal bolt
- Washer
- Hexagonal nut

8. B, C, E and G
9. **a)** X axis
   b) Y axis
   c) Z axis

Quick Test Answers
Page 17

1. **Accept any two of the following:** Dust mask; Goggles; Face visor; Apron
2. First Aid. It would be found at a first aid station or on a first aid box.
3. They are toxic and will pass into your body through your skin.
4. The spray mist will get into the air and spread around the room and could be a breathing hazard.
5. **Accept any two of the following:** Check for electrical and mechanical damage; Make sure you are wearing safety equipment; Make sure you know where the power-off button is; Check that any machine guards are in place.
6. **Accept any three suitable designers, e.g.:** Charles Rennie Mackintosh; Philippe Starck; James Dyson; Frank Lloyd Wright; Ettore Sottsass

Page 19

1. **Accept one of the following:** Look after health and safety of workforce; Use 'green' raw materials; Reduce energy use.
2. Carbon in the atmosphere causes global warming, and there is increasing awareness of the need to reduce our carbon footprint.

3. **Accept any three of the following:** Reduce the amount of material used in manufacture; Recycle materials; Re-use materials; Repair products; Refuse to accept unethical or wasteful designs; Re-think our attitude to environmental impact.
4. Re-using a material means using it again for something else, recycling it means melting it down or making it into another form before re-using it.

Page 23

1. Quality Assurance.
2. **Accept any two of the following:** Answering survey questions; Joining focus groups; Filling in questionnaires.
3. Quality Control.
4. **Accept any two of the following:** Monitor size; Colour; Taste; Electrical safety; Flammability.
5. Continuous production

Page 27

1. An outside caliper.
2. Spirit level.
3. Vernier gauge / micrometer.
4. To check that something is the correct length.

Production and Manufacture (Cont.)

Answers to Practice Questions
Page 28

1. **a)** Wear safety gloves to protect your hands from heat or chemicals.
   b) Wear safety goggles to protect eyes from dust.
   c) Wear a safety mask to make sure that you don't breathe in any harmful vapours.
2. gauntlets; face; dust; hot; in a safe area
3. **a)–b) Accept any two of the following:** To see if they have any good ideas which can be copied; To widen the range of research available; To show to a client so they can see the different options available.
4. A 6; B 4; C 2; D 1; E 3; F 5

Page 29

5. A, E and F

Woods

Quick Test Answers
Page 31

1. Because of the damage the metal head does to the wooden handle.
2. To shape metal rivet heads.
3. Soft jaws need to be used.
4.

 Corner Cramp

Page 33

1. The designer can select the most appropriate properties that will make the best product.
2. **Accept any three of the following:** Strength; Tensile strength; Brittleness; Toughness; Shear; Compressive strength; Ductility; Hardness; Work hardness; Durability; Flexibility; Elasticity; Plasticity; Malleability; Impact resistance; Electrical conductivity; Thermal conductivity; Chemical resistance.
3. Because the wood is not sustainable and deforestation leads to climate change and loss of habitat for people and animals.
4. To make our natural resources last longer and to use less energy (making new cans uses more energy than recycling cans).
5. Oil will run out and there will be disposal problems with non-recyclable plastics.

Page 39

1.

2. Medium Density Fibreboard
3. Compressed wood dust and resin glues.
4. **Accept any three of the following:** Cutting knife; Tri Square; Steel Rule; Pencil
5. To sever the wood fibres in order to make a neat job.

Page 41

1. To plane the wood up roughly to size.
2.

3. A coping saw or hole saw
4. When sawing across the grain the fibres need to be cut with a cross cut saw. When sawing along the grain they need to be chiselled with ripsaw teeth.

6. **a)** Comes up with ideas for the new product and shares them with the client; works with manufacturer to make prototype.
   b) Commissions a piece of work and is consulted about work as it progresses.
   c) Specialises in certain types of production; works with designer to produce a prototype.
   d) Anyone who buys or has access to the product.
7. B and C
8. It would be used to make a series of identical products that are made together in either small or large quantities.
9. **a)–b) In any order:** To make each stage of the process clear; To help prevent mistakes, e.g. to make sure that the varnish has not been applied before it is sanded.

Page 45

1. Increase the gluing area.
2. Dowels
3. They are used as temporary joints to hold things together while glue dries or in carpentry and general building work.
4. Knock Down.

Page 47

1. Polyvinyl acetate (PVA).
2. Epoxy resin
3.

4. **Accept any two of the following:** To prevent dirt from spoiling it; To give it a waterproof surface; To make it more attractive; To protect it from insect attack.

Answers to Practice Questions
Page 48

1. **a)–b) Accept any two of the following:** Straight-grained but knotty; Cream / pale brown in colour; Fairly strong but easy to work with; Cheap
2. **Accept any one of the following:** Outdoor furniture; Boat building; Laboratory furniture and equipment
3. **a)–b) Accept any two of the following:** Open grained; Easy to work with; Pale cream colour, easy to stain
4. B
5. A, D and E
6. A 6; B 5; C 2; D 1; E 3; F 4

Page 49

7. **a)** bevel-edged
   b) horizontal
8. **a)** PVA
   b) Synthetic resin
   c) Latex adhesive
   d) Contact adhesive
9. **a)** Medium Density Fibreboard
   b) Polyvinyl Acetate
10. **a)** B
    b) C
11. hardwood; origins; tropical rain forest; sustainable; two for one

Answers

Quick Test Answers
Page 51
1. Ferrous, non-ferrous and alloys.
2. **Accept any three of the following:** Workability; Structural strength; Appearance; Elasticity; Ductility; Malleability; Hardness and work hardness; Brittleness; Toughness; Tensile strength; Compressive strength; Resistance to rusting / tarnishing.
3. Blue
4. To soften it when it becomes work hardened.

Page 55
1. **Accept any three of the following:** Precious metal; Very ductile and malleable; Isn't affected by oxidation.
2. It is ductile and malleable, resistant to corrosion and doesn't taste or taint the food.
3. It is used to show up marking out on a metal surface.
4. A dot punch is used for marking the centre of holes or to show where to file to on a line.
A centre punch is used for enlarging the centre of holes for drilling into metal.
5. In welding, the metals being joined are melted. In soldering, a bonding alloy is melted and used to join the metals.

Page 59
1. Metal Inert Gas
2. Spot welding.
3. A twist drill.
4. Threading inside holes.
5. A die stock.

Page 61
1. Because the metal follows the shape of the forging so the grain of the metal isn't interrupted.
2. Teeth per inch
3. To set the blade, i.e. make a wider cut to reduce friction.
4. At least three.
5. A piercing saw.

Page 63
1. The top box in the pair Cope and Drag which the sand is packed into.
2. The holes that will form the pouring spout and another, the riser, to let the air out.
3. A crucible.
4. Making jewellery to produce rings and brooches.

Page 65
1. High carbon steel
2. Because they are brittle and can snap if dropped or abused.
3. **Accept any three of the following:** Rough, bastard, second cut, smooth cut, dead smooth
4. To file a corner of less than 90°
5. No finish required but can be polished or wire brushed for effect. The edge needs to be polished.

Answers to Practice Questions
Pages 66-67
1. C
2. A, D and E
3. a) Ductile and malleable, rusts when exposed to moisture.
   b) Nuts, bolts, car bodies, furniture frames, gates, girders
   c) Very strong in compression, but brittle
   d) Metalwork vices, brake discs and drums, car cylinder blocks, manhole and drain covers, machinery bases
   e) Hard, yellow metal, often cast and machined.
   f) Decorative metal work, plumbing accessories
   g) Light and anodised to protect the surface and to colour it
   h) Cooking foil, saucepans, chocolate wrappers, window frames, ladders
4. A 5; B 1; C 4; D 2; E 3; F 6
5. a) Metal Inert Gas.
   b) Tungsten Inert Gas.
   c) Teeth Per Inch.
6. a) Flat file
   b) Bastard
   c) Square file
   d) Second cut
   e) Half round file
   f) Second cut
   g) Flat file
   h) Smooth cut
7. a) B
8. b) C
9. cross; odd-leg callipers; dot punch; centre punch

Quick Test Answers
Page 71
1. Thermoplastic and thermosetting plastic
2. Thermoplastic will soften on heating and can be re-formed many times. Thermosetting plastic will not change shape and is not affected by heat once moulded.
3. Urea formaldehyde because it isn't an electrical conductor and isn't affected by heat.
4. **Accept any three of the following:** Tri square; Ruler; Fine-liner pen; Pencil.
5. It will not scratch the surface and can be wiped off easily with solvent.

Page 75
1. A wooden or metal support which guides and holds the material at the correct angle when being bent.
2. It moulds heated plastic onto the shape of former and the softened plastic is held by the vacuum until it has cooled.
3. Polyvinyl chloride.
4. Blow moulding.
5. Compression moulding.

Page 79
1. A former / mould.
2. Solvent cement or epoxy resin.
3. In a product where a change of temperature needs to be shown by a colour change, e.g. thermometer, novelty coffee cup or safe spoon for a baby.

Plastics (Cont.)

4. Cement mixed with water, sand and aggregate, which has been reinforced by the addition of steel bars.
5. In a thermostat circuit – if the temperature changes, the metal with the greater rate of expansion bends round the other, breaking or making the circuit.

Answers to Practice Questions
Page 80
1. **a)** thermosetting
   b) thermoplastic
2. C
3. D
4. **a)–b) Accept any two of the following:** Hard and brittle; Dark colour with a glossy finish; Heat resistant
5. **a)–b) Accept any two of the following:** Pipes; Bowls; Milk crates;

Buckets; Packaging; Film; Carrier bags; Toys; 'Squeezy' detergent bottles
6. **a)–b) Accept any two of the following:** Hard, Resistant to wear and tear; High melting point
7. **a)** Polyvinyl chloride
   b) High impact polystyrene
   c) High density polythene
8. **a)** File, wet and dry paper, buffing machine and brasso
   b) Cross and draw file with fine teeth; wet and dry work from coarse to fine grades; buff with compound; hand polish with a cloth
Page 81
9 A 2; B 5; C 6; D 1; E 4; F 3
10. D
11. mould; inject; female; tap

Systems and Controls

Quick Test Answers
Page 83
1. Computer numeric control
2. Used to support long pieces of material or fitted with a drill chuck for drilling holes.
3. Clockwise
4. Stepper motor
Page 87
1. Rotating / rotary, linear, reciprocating and oscillating
2. Fulcrum
3. Reciprocating or linear to rotary (or vice versa).
4. Extension, compression, radial movement and twisting.
Page 89
1. A chain doesn't stretch or slip and has a direct drive.
2. A belt will stretch allowing it to be changed easily and to absorb shock, but it may slip.
3. They use less power and are much tougher.

Answers to Practice Questions
Page 90
1. **a)** oscillating
   b) reciprocating
2. B
3. D
4. **a)** **i)–iii) Accept any three of the following:** Roughing tool; Knife tool (left and right handed); Round nosed tool; Parting tool; Boring tool
   b) **i)–iii) Accept any three of the following:** Skew chisel; Gouge; Parting chisel; Scraper
5.

It is used to transfer motion to control how fast something turns and to make lifting easier, for example CD players, washing machines and cranes.

6. **a)** How much the follower moves from the lowest to the highest position of the cam.
   b) Holds the follower in place.
   c) A rod which moves up and down on top of the cam.
Page 91
7. A 3; B 2; C 1
8. **a)** Bevel gear

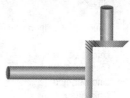

   b) Crank

   c) Crank

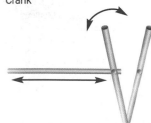

9. B

Index

Acknowledgements

P.5 ©iStockphoto.com/
Matthew Silber
P.6 ©iStockphoto.com/
Justin Welzien
P.9 ©iStockphoto.com/Franck
Boston/Joachim Angeltun
P.17 ©iStockphoto.com/Joanne
Green/Hugo Chang/
Ken Sorrie
P.23 ©iStockphoto.com/
Trevor Fisher
P.25 ©iStockphoto.com/
Brian McEntire
P.35 ©iStockphoto.com/
Bernardo Grijalva
P.37 ©iStockphoto.com/Bill Noll
P.46 ©iStockphoto.com/
Daniel R. Burch/Owen Price
P.50 ©iStockphoto.com
P.54 ©iStockphoto.com/
Christoph Ermel
P.56 ©iStockphoto.com/
Glen Jones/Owen Price

P.57 ©iStockphoto.com
P.60 ©iStockphoto.com/
Bojan Fatur
P.69 ©iStockphoto.com/
Eliza Snow/Yuriy Panyukov
P.70 ©iStockphoto.com/
David Meharey
P.82 ©iStockphoto.com/
Andris Daugovich/
Joerg Reimann
P.85 ©iStockphoto.com/
Donald Erickson
P.87 ©iStockphoto.com/
Jakub Semeniuk

The following images are reproduced
with the kind permission of Sealey.
www.sealey.co.uk
P.24 SM1304
P.26 AK9630M; S0707; S0475;
AK6086
P.30 WV175; CV100XT; 100DV;
AK6002; AK6036; AK6101

P.38 AK6086; AK6088; AK6089
P.40 SM1314; AK6093; AK6091;
CP1835VHK
P.41 GDM120BX; BHS83
P.42 AK8641; SM43
P.54 AK11150
P.55 AK9751; AK9765; AK9798
P.56 SR122
P.58 3S-4R; DF910
P.61 AK869; AK868
P.64 AK576
P.71 AK6088
P.75 AK6092

The following images are reproduced
with the kind permission of Rapid
Electronics Ltd., Severalls Lane,
Colchester, Essex CO4 5JS.
www.rapidonline.com
P.30 861382; 861416
P.40 860082
P.41 862260
P.46 861708

P.77 060768; 060772
P.78 870090; 870090; 061278

Controlled Assessment Guide
P.5 ©iStockphoto.com/
Julien Grondin
P.14 ©iStockphoto.com

All other images ©2009
Jupiterimages Corporation,
and Letts and Lonsdale.